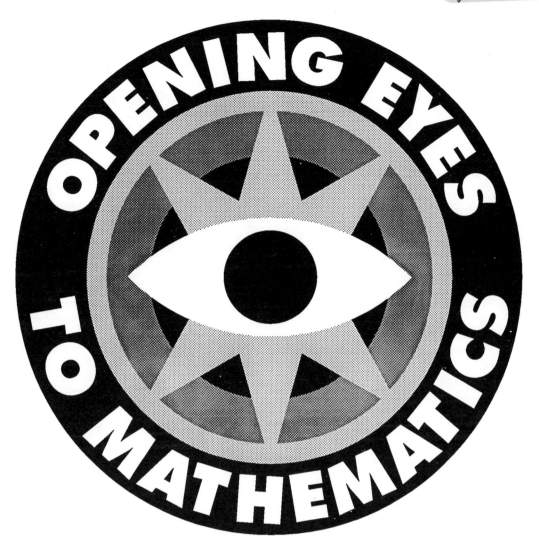

CONTACT & INSIGHT LESSONS / VOLUME 1

Debby Head
M.A., Elementary School Teacher

Libby Pollett
M.A., Elementary School Teacher

Michael J. Arcidiacono
D.Ed., The Math Learning Center

Published by The Math Learning Center
Salem, Oregon

Opening Eyes to Mathematics
Contact & Insight Lessons / Volume 1
Includes a set of blackline masters, packaged separately

Opening Eyes to Mathematics consists of
Teaching Reference Manual (Grades 3 and 4)
Contact & Insight Lessons / Volume 1 and blackline masters (Grade 3)
Contact & Insight Lessons / Volume 2 and blackline masters (Grade 3)
Lessons / Volume 3 and blackline masters (Grade 4)

Library of Congress Catalog Card Number: 91-62683

Copyright ©1991 by The Math Learning Center, P.O. Box 3226,
Salem, Oregon 97302. (503) 370-8130. All rights reserved.

The Math Learning Center grants permission to classroom teachers to
reproduce the blackline masters (packaged separately) in appropriate
quantities for their classroom use.

Prepared for publication on Macintosh Desktop Publishing system.

Printed in the United States of America by

ISBN 1-886131-31-7

Contents

Contact Lessons
page

1	Mystery Box Sorting	1
2, 3	Graphing Ourselves	3
4	Getting to Know You—Shapes	5
5	Getting to Know You—Extended Number Pattern	6
6	Modeling the 2 Counting Pattern with Rectangles—Part I	7
7	Modeling the 2 Counting Pattern with Rectangles—Part II	9
8	The 2s—Further Exploration	10
9	The 2s—Dimensions and Areas; Song	13
11	Apples—Measurement	
12	Apples—Sorting	
13	Apples—Measurement	
14	Apples— Probability and Statistics	
15	Apples—Estimation and Place Value	
16	Apples—Patterns	
17	Apples—Symmetry	
18	Apples—Money	27
19	Apples—Extended Number Pattern	28
20	Modeling the 5 Counting Pattern with Rectangles—Part I	30
21	Modeling the 5 Counting Pattern with Rectangles—Part II	32
22	Further Exploration of the 5 Counting Pattern	33
23	5s—Dimensions and Areas; Song	35
24	Mystery Box—Football; 5s—Using a Calculator	37
25	Geometry—Wind Socks!	39
26, 27	Geometry—Symmetry	40
28	Circumference—Estimation and Comparison	42
29	Volume	43
30	Volume	45
31	People Fraction Bars® *	46
32	Diagramming People Fraction Bars	49
33	Introduction to Group Fraction Bars	51
34, 35	Group Fraction Bars	52
36	Football—Geometry	
37	Football—Measurement	
38	Football—Money	
39	Football—Statistics	
40	Football—Problem Solving	
41, 42	Football—Story Problems	
43	Football—Fractions/Measurement	
44	Football—Problem Solving and Patterns	65
45–49	Exploring the 3 Counting Pattern	67
50	Mystery Box—Money; Geometry—Body Shapes	68
51	Geometry—Elastic Shapes	70

*Fraction Bars® is a registered trademark of Scott Resources, Inc.

Contact Lessons (continued)

Lesson	Topic	page
52, 53	Geometry—Shapes	72
54	Geometry—Solid Shapes	74
55	Measurement—Non-Standard vs. Standard Units of Weight	75
56	Measurement—Weight, Estimation and Graphing	77
57	Money—Estimation and Place Value	78
58	Money—Sorting and Fractions	80
59, 60	Money—Graphing	81
61, 62	Money—Congruence and Paper Folding	82
63	Money—Geometry	85
64, 65	Money—Story Problems	86
66	Mystery Box—Flags; Geometry—Symmetry	87
67, 68	Geometry—Congruence	89
69	Measurement—Volume	91
70	Spatial Relationships—Cube Covers	92
71	Measurement—Surface Area	93
72	Measurement—Surface Area	95
73	Fractions—Fraction Bar Wall Chart	97
74, 75	Fraction Bar Discoveries	98
76	Flags Sorting—Venn Diagrams	100
77	Flags—Patterns	102
78	Flags—Estimation, Measurement	104
79	Flags—Measurement	105
80, 81	Flags—Story Problems, Money	106
82	Flags—Probability	107
83	Flags—Extended Number Patterns	109
84–88	Exploring the 4 Counting Pattern	110
89	Measurement—Weight	112
90	Volume—Estimation	113
91	Geometry—Spatial Visualization	114

Insight Lessons

		page
1	Getting to Know You—People Sorting	117
2	Getting to Know You—Estimation	118
3	Getting to Know You—Probability	119
4	Getting to Know You—Money	120
5	Place Value—Base Five Pieces	121
6	Place Value—Totaling Units Using Base Five Pieces	122
7	Place Value—Comparing Collections	124
8	Place Value—Representing a Number by Different Collections	126
9	Place Value—Identifying the Minimal Collection	128
10	Place Value—Forming Minimal Collections	130
11	Place Value— Forming Minimal Collections	132
12	Place Value—Addition and Subtraction	133
13	Place Value—Making Base Two Pieces	134
14	Place Value— Making Base Three Pieces	136
15	Place Value—Making Base Eight Pieces (Optional)	138
16	Place Value—Base Ten Awareness	140
17	Place Value—Forming Collections	142
18	Place Value—Ways to Represent a Number	143
19	Place Value—Using Base Ten Grid Paper	145
20	Place Value—Drawings	147
21	Place Value—Expanded Forms	149
22	Place Value—More and Less	151
23	Place Value—More-and-Less Diagrams	152
24	Place Value—Comparing Numbers	153
25	Place Value—Up and Back Again	154
26	Communication of Process	155
27	Place Value—Sketches	157
28	Addition/Subtraction Operations, Number Sense & Process	159
29	Place Value—Out of Bounds	161
30	Place Value—Rounding	162
31	Place Value—Rounding	163
32	Place Value—In the Mind's Eye	164
33	Exploring the Difference Between Two Numbers	165
34	Difference—Diagrams and Sketches	167
35	Big Difference/Little Difference	170
36	The Great Race	172
37	Estimating and Calculating Differences	173
38	Differences—Teamwork	175
39	Differences—Back Me Up	176
40	Differences—Rounding	178
41	Calculator Differences	179
42	Differences—"Tell It All" Books	180
43	Same Differences	182

Insight Lessons (continued)

page

44	Difference Duel	184
45	Modeling Addition Stories	185
46	Mine, Yours, Ours—Modeling Addition	190
47	Practicing Addition	192
48	Addition—Phoney Numbers	193
49	Jersey Day—Multiple Addends	195
50	Number Card Addition	196
51	Subtraction Take-Away Stories	197
52	Strategy—A Game of Take Away	201
53	Number Card Take Away	203
54	Stories That Mix Addition and Subtraction	206
55	Addition and Subtraction—Teamwork	207
56	Addition and Subtraction—Tell It All Story Books	209
57	Relating Addition and Subtraction—Building Partitions	210
58	Partition Challenge	213
59	Ways to Partition 100	214
60	Back Me Up—Mental Pictures of Addition and Subtraction	216
61	Take A Long Hard Look—Mental Arithmetic and Calculating Options	218
62	Estimation, Rounding and Calculators	221
63	Number Card Tic-Tac-Toe—Mental Arithmetic and Problem Solving	223
64	How'd Ja Do It?	225
65	Do It All	227
66	Addition and Subtraction in the Mind's Eye	229
67	Toy Store Spree	230

Index for Teaching Reference Manual and Lessons: Volumes 1, 2 and 3 231

Introduction to Lessons

Opening Eyes to Mathematics consists of three types of lessons:

- Calendar Extravaganza
- Contact Lessons
- Insight Lessons

In general, these will require a total of 80 to 90 minutes of class time each day, perhaps as suggested here:

      8–8:20   Calendar Activity
   8:20–10:30   Language Arts
   10:30–11   Contact Lesson
   11–11:30   Insight Lesson

Calendar Extravaganza

The Calendar Extravaganza (included in the Teaching Reference Manual) consists of 20- to 30-minute explorations of several mathematical concepts. These explorations are related to each day's date and often connect mathematics to other subject areas. We urge you to use teacher choice in determining which components to emphasize each month. In that way you will create an Extravaganza that best meets the needs of your children. We are confident that you and your children will enjoy these daily activities.

Contact Lessons

Contact lessons make contact with fractions, measurement, geometry, probability, data analysis and additional numeration activities. Many of these lessons are also set in the context of monthly themes (such as football, apples and stamps) that link mathematics to the world of children.

Contact lessons are organized in a spiraling fashion so that, as children mature and grow in confidence, they explore concepts again and again throughout the year. There are 185 lessons; each can typically be conducted in 30 minutes. As with Insight lessons, however, we encourage you to exercise teacher choice when using them. You may wish to conduct them as written or alter (or even replace) them to suit the needs and interests of your children. For example, if your children are especially interested in learning about magnets, you might choose magnets to be one of your monthly themes.

Insight Lessons

Insight lessons provide daily experiences with numeration topics and help children strengthen their sense of numbers. During Insight lessons, children learn about place value, decimals, addition, subtraction, multiplication and division. They also develop their ability to make reasonable estimates, choose appropriate calculating options and solve problems.

To assist you with planning, we have sequenced the Insight lessons and suggest the range of days it might take to complete each one. You may change or even skip some lessons; others you might decide to extend.

The Insight lessons are intended to be a roadmap that provides general directions for your children's journey through their study of numeration. We urge you to trust your professional judgement and intuition, adjusting the path of this journey so that it is appropriate for your children, taking detours or expressways as needed.

Most Insight and Contact lessons are written in a brief style. Throughout the lessons, there are references to chapters in the Teaching Reference Manual of this program—please refer to these chapters as you prepare to teach. Where appropriate, the lessons contain examples, discussion questions and possible responses from children. They also include suggestions for homework, assessement and journal writing. Since neither lesson is directly dependent upon the other, the Insight lesson does not need to immediately follow the Contact lesson in the daily schedule.

The lessons encourage "child choice" of materials and methods for solving problems, as well as opportunities for children to participate in mathematical discussions. In many lessons, for example, children demonstrate their thinking in a show-and-tell manner or write about their discoveries. This helps children strengthen their understanding of mathematical concepts and learn from one another.

The spirit described in Chapter 1 of the Teaching Reference Manual is the basis for these lessons. We urge you to jump in and give the lessons a whirl—we think you will find them exciting! Be ready for surprises and to learn along with your children! Why, just this week, one of our children connected the 7 counting pattern (as it appeared on a hundreds matrix) to the game of chess. A quick trip to the library helped everyone (including us) learn more about this connection. What fun!

Contact Lessons
1 – 91

1 Mystery Box Sorting

You Will Need
- Chapter 2, Sorting
- a mystery box that contains an apple
- chart paper
- markers

Your Lesson

We like to introduce each Contact theme (except for Getting to Know You) by placing an object representing it in a "Mystery Box". The children then ask yes-or-no questions about what is in the box, and the information obtained is recorded on chart paper. The number of questions entertained on a given day is limited, so, if the quota is reached before the object is identified, the children must wait until the next class to try again. Start this activity a few days before the other lessons related to the theme.

The mystery item this time is an apple! The Apples theme begins with Lesson 11 and we are sure it will provide you and your children with many ap-"peel"-ing mathematical experiences! After all, an apple a day keeps the doctor away, and is certainly the way to a teacher's heart. This theme is sure to open the hearts of children to a love of mathematics as well!

Let the questioning begin! You might keep track of them by drawing a 12- to 15-part design on the board. As each question is answered, erase one of the parts. When all parts are gone, the questioning stops for the day.

1 Mystery Box Sorting (continued)

Sometimes children find it hard to ask questions about the attributes of the object. Instead they may try to guess the contents directly by asking, "Is it a picture?" "Is it a block?" They may require examples of helpful questions, such as "Is it small?" or "Can it be found in school?" They may also need to be reminded that even "No" answers provide useful information.

One way to help children improve their questioning and reasoning skills is to periodically review the clues that have been gathered and to try posing new questions based on them. Another way is to limit the number of questions that can be asked during the lesson. You might note which questions are *guesses* of the contents and which refer to *attributes* of the object.

Teacher Tips

The children enjoy it when some theatrics are part of the lesson. For example, we begin by pretending that the box contains the secret of the year and by humming the tune, "I know something you don't know, he-he-he-he-ha-ha! I know what is in this box, he-he-he-he-ha-ha! Bet you can't guess it! He-he-he-he-ha-ha!" Naturally the children want to guess!

If the questions for the day are exhausted, we suggest keeping the box and the information chart on display throughout the day for reference purposes. This, along with the children's curiosity, will usually prompt continued discussions both at school and at home about what is in the box. (We've even had parents ask us in the grocery, "Just what *is* in that box?", especially if the questioning has gone on for a number of days.) Some time during your day tomorrow, review today's questions and draw a new figure to be erased bit by bit. Have fun! It can be done just before lunch or whenever 5 minutes appear in your schedule!

Continue as many days as needed. When the box is finally opened, you will find exploring mathematics with apples as easy as apple pie!

Additional Ideas

2, 3 Graphing Ourselves

You Will Need
- Chapter 2, Sorting; Chapter 10, Data Analysis and Graphing
- graphing materials to go with the style of graphs you choose

Your Lesson

The Getting to Know You theme sets the stage for a year of friendships. Try these lessons and enjoy getting to know your children. And by the way, don't leave yourself out; they'll enjoy getting to know you, too!

Ask the children to share information about themselves and help them prepare graphs that display it. Once the graphs have been constructed, invite the children to record observations about them. Here are some examples.

Bar Graphs

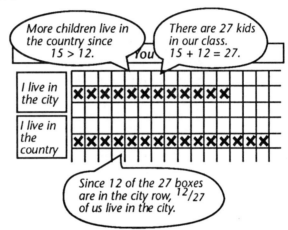

Venn Diagrams

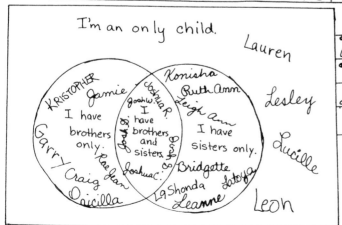

2, 3 Graphing Ourselves (continued)

Pie Graphs

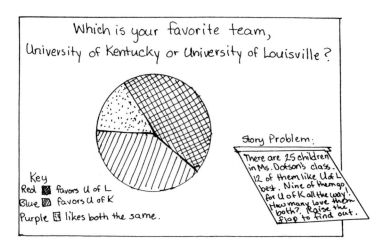

Teacher Tips

You may wish to precede these graphing experiences by having the children represent the information being displayed. For example, they can create "real" graphs by arranging themselves into the columns of a bar graph or into the regions of a Venn Diagram.

Students enjoy deciding how to record data. For example, instead of coloring rectangles to make a bar graph, they may suggest using checks, initials or smiley faces.

Have children discuss the graphs at the first Open House so their parents also "get to know" the class.

This could be adapted for use with a computer data base. Go for it!

Additional Ideas

4 Getting to Know You—Shapes

You Will Need
- art materials for each child (paper, markers, pencils, crayons, etc.)

Your Lesson

What shapes can we find on our faces?

Have each child sketch the face of a friend, noting the basic geometric shapes they see. Does the face have an oval outline? Is there a resemblance between the nose and a triangle? Can any circles be observed?

You might introduce this activity by looking for shapes that are present in magazine pictures or by discussing how artists look for certain shapes in the faces of their models as they draw. The children will also enjoy watching you sketch a likeness of your own model (a principal or a counselor, for example).

Teacher Tips

We like to have the children color their drawings and then display them on the bulletin board. The display is titled, "Our Class Is Really Shaping Up!"

You might consider inviting a make-up consultant to discuss different facial shapes with the class.

We were pleasantly surprised how much the children's drawings looked liked their friends. We believe this was largely due to having spent time modeling and analyzing shapes prior to turning them loose.

Journal Writing

Complete the following sentence: "Today, I enjoyed _____."

Homework

Have your children examine their reflections in a mirror. What shapes are noticed? Ask each child to draw a self-portrait. How did the shape examination help?

5 Getting to Know You—Extended Number Pattern

You Will Need
- Chapter 3, Patterns; Chapter 6, Multiplication and Division
- large paper for background of display
  for each child
- 11 × 18 art paper
- pencils, crayons, scissors, glue

Your Lesson

How many feet are present in your room?

Invite the children to suggest strategies for answering this question. Some might suggest counting by 2s, while others might want to count 10 feet for every group of 5 people that can be formed and then add 2 for each person left over. Of course, one could also multiply the number of people in the room by 2.

Have the children trace around their shoes and decorate them appropriately. Create a class display and label each set as shown.

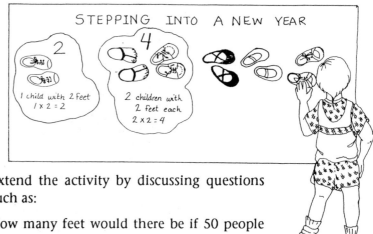

Extend the activity by discussing questions such as:

How many feet would there be if 50 people were present?

How about if 1,000 people were in the room?

Suppose there are 78 feet altogether. How many people are in the room?

Could there ever be a total of 45 feet in the room? Why or why not?

Teacher Tips

The activity assumes 2 feet for each person in the room. Under this assumption, the total number of feet must be even. This is a wonderful way to introduce or review even and odd numbers.

Journal Writing

If there are 95 people in a room, how many feet would there be? Describe how you would answer this question. How do you feel about your answer?

6 Modeling the 2 Counting Pattern with Rectangles—Part I

You Will Need
- Chapter 3, Patterns; Chapter 6, Multiplication and Division
- colored squares for the overhead and markers *for each child or group*
- units from the base ten counting pieces
- linear units
- ten 3 × 5 index cards or pieces of paper

Your Lesson

Contact Lesson 5 was the first of six lessons that focus on the counting progression: 2, 4, 6, 8, 10, etc. The goal of the lessons is to help your children think about this progression in terms of pictures, patterns and song. All of the senses are brought into play. Today, this counting pattern is modeled by rectangles formed from the units of the base ten counting pieces.

Build the following rectangles at the overhead and have the children work in teams of two to do the same at their desks. As the children work, facilitate and observe by cruising from group to group. Have a friendly chat about the work being done.

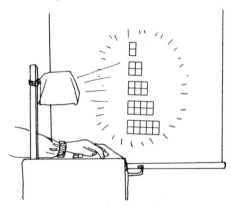

Ask them to continue building rectangles of this sort. Since these are intended to be a model for the 2 counting pattern, each rectangle should have 2 rows. Each should also have 2 more units than the one before it

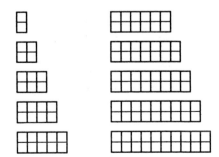

6 Modeling the 2 Counting Pattern...(continued)

Discuss the areas of these rectangles. These areas show the counting progression: 2, 4, 6, 8, 10, 12, 14, 16, 18, 20. Be sure to note this connection.

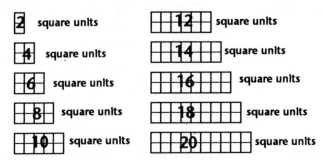

Now have the children use linear units to show the dimensions of these rectangles. Discuss the patterns of dimensions: one dimension is always 2, while the other increases by 1 unit each time.

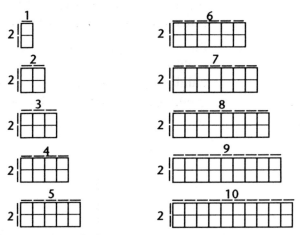

Complete the lesson by having the children label the dimensions and area of each rectangle on an index card.

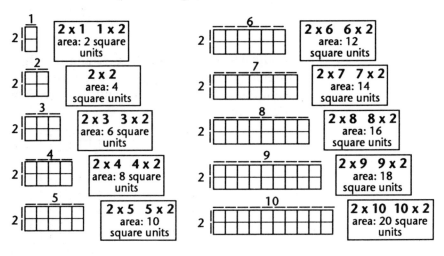

Additional Ideas

8
CONTACT LESSONS

7 Modeling the 2 Counting Pattern with Rectangles—Part II

You Will Need
- Chapter 3, Patterns; Chapter 6, Multiplication and Division
- base ten counting piece units
- several sheets of centimeter grid paper (Blackline 25)
- 18 × 24 construction paper and glue for each child

Your Lesson

Have children recall the rectangles built during Lesson 6. In pairs, have the children build the progression of rectangles once more using the units from the base ten counting pieces. When they are finished, ask them to prepare a display by cutting the rectangles from graph paper and gluing them to construction paper. The rectangles should be arranged in area from smallest to largest, as shown below. The dimensions and areas should also be indicated. Save these displays for the next lesson.

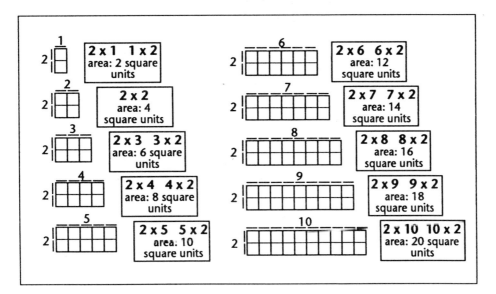

Encourage the children to share their work with a friend and practice the 2 counting pattern.

Teacher Tips

Encourage the children to work cooperatively on this activity. Save these displays for tomorrow's lesson.

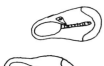

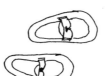

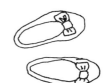

8 The 2s—Further Exploration

You Will Need
- Chapter 3, Patterns; Chapter 6, Multiplication and Division
- the displays of rectangles from Lesson 7
- colored squares for the overhead and a set of linear units *for each child*
- base ten counting piece units
- large laminated hundreds matrices (Blackline 38)
- individual hundreds matrix (Blackline 27)
- crayons

Your Lesson

Are your children beginning to "see" mental pictures of the rectangles that show the 2 counting pattern? Before beginning today's lesson, practice just a bit.

TEACHER *Close your eyes— and try to picture a 2 by 4 rectangle. What is the area?*

CHILD *I see a 2 by 4—that's 2 rows of 4—and that makes 8 square units in the area of that rectangle.*

The teacher displays the rectangle at the overhead, and invites a volunteer to show the dimensions of the rectangle with line units.

TEACHER *Thank you. Now try to picture a rectangle that has 12 square units. One of the dimensions is 2. Can you picture this rectangle? Raise your hand when you think you can "see" the other dimension.*

CHILD *I see it! The rectangle has 2 rows of 6 square units each. The dimensions are 2 on one side and 6 on the other side! The other dimension is 6!*

8 The 2s—Further Exploration (continued)

Return the displays from Lesson 7 to be used as a model and lead your children in more exploration of the 2 counting pattern. Some snapping and clapping would be appropriate here.

TEACHER *I am going to snap and clap the 2 counting pattern. I will snap on all numbers that are not in this pattern, but when I come to one that is, I will clap. Join me when you "feel" the pattern.*

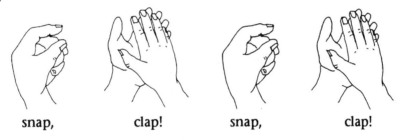

snap,   clap!   snap,   clap!

When the children have found the rhythm to this, ask them to count at the same time, whispering on all snapped numbers and *yelling* on all clapped ones. Discuss the fact that the clapped numbers will always be even and correspond to the areas of their rectangles.

The 2 counting pattern can also be depicted on a hundreds matrix. Ask your children to cover with base ten units all the numbers in this pattern through 100 on such a matrix.

1		3		5		7		9	
11		13		15		17		19	
21		23		25		27		29	
31		33		35		37		39	
41		43		45		47		49	
51		53		55		57		59	
61		63		65		67		69	
71		73		75		77		79	
81		83		85		87		89	
91		93		95		97		99	1

8 The 2s—Further Exploration (continued)

What patterns do they notice in the arrangement of squares on the matrix? If the matrix was extended would 102 be covered? How about 403 or 2,000? Why or why not?

Complete the lesson by having each child shade in the 2 pattern on a small hundreds matrix to keep as a permanent record. These can be glued to their display of rectangles.

Teacher Tips

Now for bragging rights! Display the finished displays all over—in the room, in the hall, wherever!

There are many patterns to be noticed when the multiples of 2s are shaded on a hundreds matrix. For example, only 5 of the columns are shaded and these contain all even numbers. Observations such as these, together with your children's mental perceptions of even and odd numbers, are helpful for predicting if larger numbers will fit the pattern. Calculators are also useful here (see Lesson 10).

Journal Writing

This is a good time for the children to do some writing. Ask them to summarize the day's events, explain some of the ways they pictured the 2 counting pattern and what they have learned about it.

Have children write a letter to their parents describing what they noticed about the 2 counting pattern!

Additional Ideas

9 The 2s—Dimensions and Areas; Song

You Will Need
- Chapter 3, Pattern; Chapter 6, Multiplication and Division
- The 10 multiplication/division discussion cards that show rectangles with a dimension of 2
- number pattern song cassette tape, *Musical Array-ngements*

Your Lesson

As you did at the start of Lesson 8, provide some time for the children to mentally picture rectangles that have a dimension of 2. Can they "see" these rectangles and identify their dimensions and areas? Ask questions such as:

Can you picture a 2 by 7 rectangle? What is its area?

Is it possible to arrange 50 square units into a rectangle that has 2 rows? If so, what would such a rectangle look like? If not, why not?

Can you give an example of a number of units that could not be arranged into a rectangle that has 2 rows?

What are some large numbers that are part of the 2 counting pattern?

Ask ten volunteers to come forward and give each of them a discussion card. Challenge them to arrange themselves so that the cards are in order from smallest area to largest. Then, beginning with the child that has the card with the smallest area, have each child hold up their card and name their rectangle.

9 The 2s—Dimensions and Areas; Song (continued)

Repeat this process, only this time have each name the area of their rectangle.

Go through the areas once more with everyone joining the recitation. Try putting in a little rhythm and pep as the number pattern is read.

Now for the song, "Two's Shoes", that is on the cassette. Sing this song for the class as the cards are again raised in sequence. Invite your children to sing along when they are ready.

Ask each child to hand their card to a friend, stating the dimensions and area of the rectangle. Have the new card holders line up in sequence and begin the singing and swaying all over again.

Invite your children to make some predictions about the cards. If the pattern of cards were continued, what would the next rectangle look like? What would its area be? How about the rectangle on the 30th card—can someone describe its dimensions and area? How about the rectangle on the 100th card? etc.

Additional Ideas

10 The 2s—Exploring with a Calculator

You Will Need
- Chapter 3, Patterns; Chapter 6, Multiplication and Division
- calculator for each child
- number pattern song cassette tape, *Musical Array-ngements*

Your Lesson

Has "Two's Shoes" hit the top ten with your children? Is the little tune resurfacing repeatedly? Sing the song with your children and write the 2 counting pattern on the chalkboard.

2	12	22
4	14	24
6	16	26
8	18	28
10	20	30

Ask them to extend this pattern by using a calculator. Challenge your children to record the number pattern as high as possible, stopping occasionally to loop digits that repeat. Encourage them to compare their record with that of a friend.

Here are some questions to be considered as they work:

What patterns can be observed in the digits of these numbers. Do they notice, for example, the 2-4-6-8-0 repetition that occurs in the units' digit?

How can one predict if a number will appear in the 2 counting pattern? What about numbers like 124? 465? 2,308? etc.

How many 2s make 40? Make a prediction and check it with a calculator, perhaps by using repeated addition. Do the same for other numbers: How many 2s make 150? 240? etc.

Can you picture a rectangle that has 2 rows with each row having 21 square units? How many square units are there altogether?

Imagine a rectangle that contains 288 square units and one dimension of 2. What are some ways to determine the other dimension? How can a calculator help here?

End by singing "Two's Shoes" once more.

Teacher Tips

Ask the children to discuss the ways they were able to extend the 2 counting pattern. Did the song help? Did any think of pictures of rectangles that have a dimension of 2? Did any feel the need for a calculator?

As your children learn additional songs in later lessons, they will love to revisit old favorites such as "Two's Shoes."

Additional Ideas

11 Apples—Measurement

You Will Need
- Chapter 7, Measurement
- an apple placed in a ziplock bag
- a long piece of adding machine tape
- several balance scales
- a kitchen scale
- apples of various sizes—one for each child
- optional: Observation Record Sheet (Blackline 81); see Teacher Tips

Your Lesson

How long does it take an apple to rot? What happens to its weight and appearance in the process?

In this lesson, your children will begin an experiment to explore these questions. Sounds like fun, doesn't it?

Divide the class into groups of four or five children, ask each group to order its apples from lightest to heaviest. They are to examine their apples and make predictions about the required order before doing any weighing. When this is done, have the children weigh the apples against each other and compare their predictions with the actual order of weight. Ask the groups to discuss or write about their results. How close were their predictions? Was there anything about their apples (such as type or circumferences) that influenced their predictions?

Now the real experiment can begin! Discuss the term "biodegradable" with your class. What are some examples of things that are biodegradable and of objects that are not? In which category is an apple?

Have a volunteer weigh an apple that has been placed in a ziplock bag as a second person records the date and the weight and condition of the apple on adding machine tape.

11 Apples—Measurement (continued)

Repeat this process daily, using the tape as a time line, until the apple is so rotten you can't stand it. Ask the children to keep a written log of their observations, describing such things as the color, texture and weight of the apple.

When the apple has rotted, discuss such questions as:

How many days passed before a change was noticed in the apple?

What happened to the weight of the apple as it began to rot?

How long did it take for the apple to rot? Would all apples require the same amount of time to rot? Why?

If this apple had been thrown out as litter, would it be considered biodegradable or not? What other living things might follow the same pattern of cellular breakdown as an apple?

Teacher Tips

If it is difficult to find more than one scale, the first part of this lesson can be conducted as a whole group activity.

You may wish to discuss the old saying "one bad apple spoils the whole bunch". What is its meaning? When picking a bushel of apples to store in the cellar, what would the pioneer families do with bruised apples? Why? Many times people use this saying when referring to other things besides apples. Can any of your children apply the saying to life situations?

You may wish to have the children organize their thoughts on an Observation Record Sheet. If this is the first time the children have used such a record, model the record-keeping on a large class-size version (use chart paper or on your classroom computer).

Additional Ideas

12 Apples—Sorting

You Will Need
- Chapter 2, Sorting
- an apple for each child

Your Lesson

What features do the children notice about their apples? Choose a volunteer to describe an attribute of their apple. Ask the other children to examine their apples for that same attribute, tallying the number of apples with the given attribute in a way all can see. In the illustration below we have begun by tallying the number of children whose apples had stems attached.

Additional apples are added to the display as you continue repeating this procedure using other attributes suggested by the children. Some that might be mentioned are color, worminess or size.

Allow a few minutes for the children to make observations about this activity. For example, how many people had apples that fit in every category? Which attribute was seen least often? Which was observed most often?

Teacher Tips

A colorful way to record results is to use apple shapes and leaves cut from construction paper, as shown above.

Or, have the children prepare graphs of the results of this sorting.

Additional Ideas

13 Apples—Measurement

You Will Need
- Chapter 7, Measurement
- an apple for each child
- hex-a-link cubes
- chart paper

Your Lesson

Suppose the children in your room were to place their apples in a straight line. About how many hex-a-links would be needed to create a line that is just as long? Conduct the following experiment to find out.

Divide the class into small groups and have each group create a chain of hex-a-links that is about as long as a train formed with its apples. Ask each group to then use the length of its apple train to estimate that of the entire class, as illustrated below.

Since there are 25 apples in our class, there will be about 6 groups of 4 apples. Our train of 4 apples is about as long as 10 hex-a-links. So we think it will take a little more than 60 hex-a-links if everyone's apples make a train.

There are 25 apples in the class... That's about 8 groups of 3 each. We needed 7 hex-a-links for our 3 apples. So we added eight 7s together and think it will take about 56 hex-a-links for the whole train.

Have each team record its estimate on chart paper. As a class, discuss these estimates. What is the range of answers? What might explain any differences that are observed? Can the children agree on a "representative" estimate that seems reasonable?

The length of a class apple train can now be compared to a chain of hex-a-links that is about as long. How close were the group estimates? How close was the "representative" estimate?

13 Apples—Measurement (continued)

Teacher Tips

In all likelihood, the team estimates will be different due to variation in apple and group size. It is important that these differences be discussed and that groups have a chance to explain their thinking. This is also a nice opportunity to discuss that measurements, by their nature, will always be inexact to some degree.

Have the groups write their estimates on post-its that can be displayed. How many of the estimates are even? How many are odd? What is the range of estimates? etc. Seize the teachable moment, moving the post-its as appropriate.

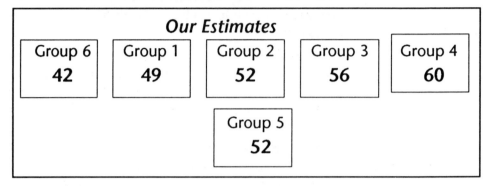

Give children a chance to agree on a representative estimate for the length of the class' train. Often an average is used for this purpose; thus this activity anticipates the average lessons that occur later in the program.

As a reminder: the edge of a hex-a-link is the unit of length in this activity and not the hex-a-link itself.

This lesson may be repeated using other measuring devices such as popsicle sticks, paper clips or rulers.

Additional Ideas

14 Apples—Probability and Statistics

You Will Need
- Chapter 9, Probability
- a transparency of Blackline 101 (9-Section Spinner) *for each group of four children*
- a spinner prepared from Blackline 101
- a copy of Blackline 58 (Apple Probability)
- (optional) Blackline 81 (Observation Record Sheet)

Your Lesson

Display your 9-section spinner and explain to the children that the letters a-p-p-l-e-s-e-e-d are to be written in the sections—one letter per section. Ask them to determine the placement of the letters. They may suggest a random arrangement such as this:

Have each group of four label its spinner in the same way.

Discuss the following questions:

If you make 20 spins, which letter do you think will be spun most often? Which letter do you think will come up the fewest number of times?

Once their predictions have been shared, have the groups make 20 spins and construct a bar graph of the results. Ask the groups to present their results to the class.

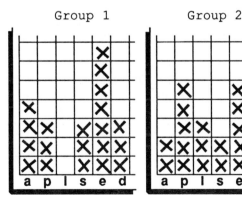

*Group 1: We spun **e** the most, just as we guessed. We didn't think **a** would come up more than **p**, since **p** was on 2 sections, but it did. We didn't get any **l**'s which was a surprise.*

*Group 2: We spun **p** and **e** the same number of times. We thought we'd get more **e**'s. The other letters came up about the same amount.*

14 Apples—Probability and Statistics (continued)

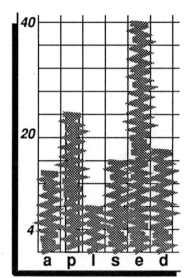

Letter	Number of Spins
a	14
p	24
l	8
s	16
e	40
d	18

When all are ready, pool the results, perhaps as shown here, and then address the questions listed below.

How closely does the data compare with the predictions? What might explain any differences?

What might occur if the experiment were repeated or if more spins were made?

Is it possible that one of the single letters could turn up most often? Is it probable that this will happen?

What fraction of the results were **a**'s? What fraction were **e**'s

What fraction of the results were **q**'s? Why?

How do the results from each group compare to the pooled results of the entire class?

Does each letter on the spinner have an equal chance of being spun? Why or why not? If not, how could the experiment be changed to give each letter an equal chance to be spun?

Teacher Tips

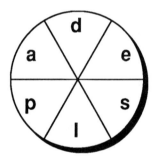

It is natural to expect the **e** and **p** to turn up most frequently. Theoretically, if the spins are random, the chances of landing on **e** and **p** are 3/9 and 2/9 respectively. If this doesn't happen, perhaps not enough spins were made or the manner of spinning was not random. Also, because 20 is a fairly small number of spins, the pooled results may be different than those of some of the groups. Theoretically, the combined totals of all the groups should be in closer agreement with what might be expected in this experiment.

The children may have various ideas for making the outcomes of this experiment equally likely. One might be to use a spinner that has six sections of the same size and to place one of the letters **a**, **p**, **l**, **e**, **s** or **d** in each section, respectively.

Another is to adjust the original spinner so that the amount of area for each letter is the same.

Consider using the Observation Record Sheet (Blackline 81). Pairs can record the appropriate information.

Additional Ideas

15 Apples—Estimation and Place Value

You Will Need
- one box of apple-flavored cereal
- a large glass bowl
- (optional) copies of Blackline 81 for each child

Your Lesson

Empty the contents of a box of apple-flavored cereal pieces into a glass bowl and pose the following problem to the class:

What are some ways we can estimate the number of whole cereal pieces in the bowl? Let children make suggestions. Also be ready to present some of your own. Some possibilities:

Make a guess after simply looking at the bowl for a period of time.

Grab a handful of cereal from the bowl and count the number of whole pieces you get. Multiply this by the approximate number of handfuls in the bowl.

Use a small cup to scoop cereal from the bowl and count the number of whole pieces. Multiply this by the approximate number of cups that could be scooped from the bowl.

Estimate the number of whole pieces that are in the top layer of the bowl. Multiply this by the number of layers that appear to be in the bowl, adjusting as best you can for the shape of the bowl.

Determine an estimate according to each suggestion made. How do the children feel about the strategies? Do they notice any problems in applying them? Which strategy seems most appropriate?

Have the children determine the actual number of cereal pieces. Teams of four can be given shares to count using groups of 10s, 100s, and (if necessary) 1,000s. Portion cups, styrofoam bowls and paper plates are helpful. The collections from each team can be combined, regrouping as needed, to arrive at a final total.

Teacher Tips

The primary purpose of this lesson is to help children develop a helpful frame of reference for estimating the number of cereal pieces. It is important to discuss the usefulness of each strategy that is suggested, and for the children to think about which one(s) they would have confidence in using.

It would be appropriate to use a calculator throughout this activity. In the suggested strategies, multiplications can be replaced by repeated addition, if necessary.

You may wish to have your children record their observations on Blackline 81.

16 Apples—Patterns

You Will Need
- Chapter 3, Patterns
- a large handful of apple-flavored cereal pieces for yourself and for each child

Your Lesson

On the overhead, with your cereal form the first 3 trains of the sequence shown below. Have the children do the same at their desks.

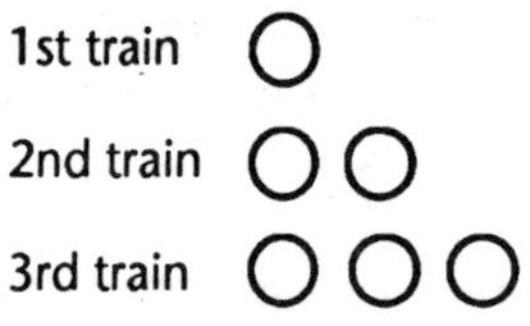

1st train ◯

2nd train ◯◯

3rd train ◯◯◯

Tell the children that you see a pattern in these trains and that you have a picture of the 4th train in mind. Ask if anyone would like to predict your pattern and build the 4th train (see first Teacher Tip).

4th train ◯◯◯◯

Challenge the students to visualize what the 10th train in the sequence would look like. Ask a volunteer to build this train without building those in between and to discuss the reason for their answer.

10th train ◯◯◯◯◯◯◯◯◯◯

What will the 23rd train look like? How about the 40th?

Now show the children another sequence for a cereal train.

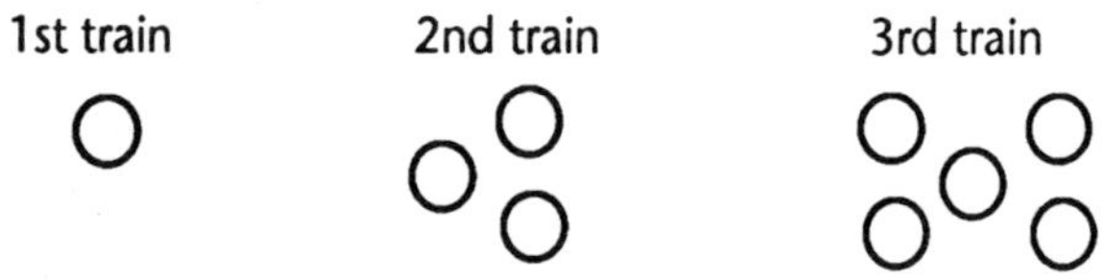

1st train   2nd train   3rd train

Ask volunteers to show the 4th and 5th trains (see first Teacher Tip).

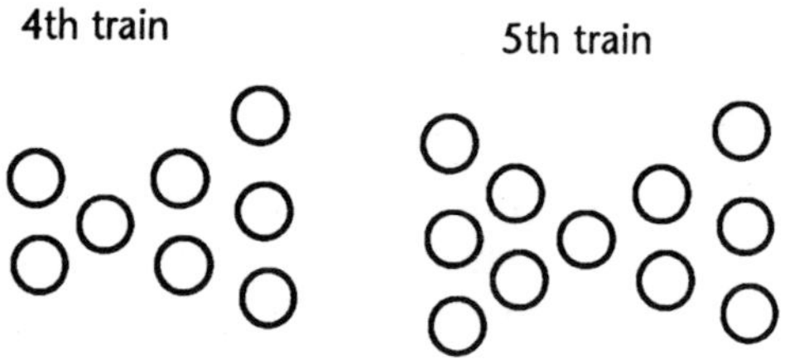

4th train   5th train

16 Apples—Patterns (continued)

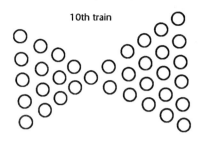
10th train

As before, ask the children to visualize the 10th train and have a volunteer build their "picture" at the overhead (see illustration).

Have the children discuss the reasons why they think the 10th train looks as it does (see second Teacher Tip.) Here are some possible responses:

"I see a pattern! Each side adds 1 more than before, with the right side increasing first."

"A bow tie shape is formed by making each row 1 piece bigger than the one beside it."

5 4 3 2 1 2 3 4 5 6

"One is added on each side to make the triangle expand by 1."

As time permits, you (or the children) can create additional patterns to explore. Some other possibilities:

	1st train	2nd train	3rd train	4th train	5th train	6th train
a)	O	O O	O O O	O O O O	O O O O	O O O O O
b)	O	O O	O O O O O	O O O O O O	O O O O O O O	O O O O O O O O
c)	O O O	O O O O	O O O O O O	O O O O O O O	O O O O O O O O O	O O O O O O O O O O O

Teacher Tips

The illustrations above display a particular pattern—the one *you* have in mind. If a child suggests some other pattern, you can acknowledge the idea and decide if you wish to explore that pattern rather than your own.

It is important that children share their thinking here. The intent is to have them form a mental image of the 10th train in each sequence and to share their reasons for their images. This may help them realize that there is often more than one way of visualizing the same arrangement.

Note that a variation of this lesson is presented in Chapter 5, Addition and Subtraction (pages 38–39). This variation examines the star arrangement formed by the seeds of an apple. This arrangement is also used to motivate the 5 counting pattern in Contact Lesson 19.

17 Apples—Symmetry

You Will Need
- Chapter 11, Geometry
- white or manila paper and crayons for each child

Your Lesson

An apple is a wonderful example of an object that has a line of symmetry. Ask your children to imagine slicing an apple into two equal parts as shown here.

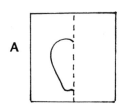
A

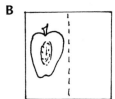
B

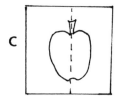
C

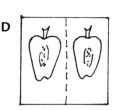
D

Now have them fold their paper and draw a side or front view of one of the parts on one side of the fold. Have them use crayons and color darkly. For example, they might draw something like illustration A or B.

Once the apple half has been completely and darkly colored, have the children refold their papers and rub them with the side of a pencil or their fists—rubbing hard enough to transfer the color to the other half of the paper, but without tearing the paper. When the paper is unfolded again, they should see a reflection of what they drew (see illustration C and D).

Invite the children to make observations about their drawings. Some questions for discussion: What role is played by the fold in this activity? How are the apple halves positioned in relation to the fold? What are other examples of objects that have a line of symmetry?

The following points can be noted in this discussion:

The fold acts as a line of symmetry and can be so identified.

Because of the symmetry present, the apple halves are mirror images of each other. The fold is exactly between the two halves.

Ask the children to identify other objects that are symmetrical. Rectangles, rainbows and human bodies are good examples. You might have pictures of others to show.

Teacher Tips

Here is another art activity that involves line symmetry. Have children blow tempera paint through a straw on the paper and then fold the paper to create a symmetrical splash! No doubt you've done it or seen it done, but did you ever relate it to the line of symmetry?

Additional Ideas

18 Apples—Money

You Will Need
- several grocery ads from the newspaper
- scissors and glue for each child
- a large sheet of art paper
- several copies Blackline 80 (Money) for every four children
- several money feely boxes

Your Lesson

In groups of four, have the children cut out an advertisement of an apple product and then use coins from their feely boxes to show the price in several ways. Solutions may be recorded on a money blackline and displayed with the ad on art paper.

As time permits, groups can repeat the process with other ads.

Discuss the displayed products. For example: Was the same product found by two different groups? If so, how did the prices compare? What might explain any differences in price?

Additional Ideas

19 Apples—Extended Number Pattern

You Will Need
- Chapter 3, Patterns; Chapter 6, Multiplication and Division
- an apple and a sharp knife
- a 4 × 4 sheet of white paper, crayons, scissors for each child
- a large sheet of bulletin board paper for display and glue

Your Lesson

If you cut an apple horizontally and look at the inside of one of the parts, what type of design will you see? How many points will it have? What type of seashell will it look like?

Answer these questions by cutting an apple as shown.

Have the children recreate the 5-point design by coloring a circular shape cut from paper. Glue some of these shapes on the bulletin board paper to create the groups shown below and identify the multiples of 5 points seen in each group.

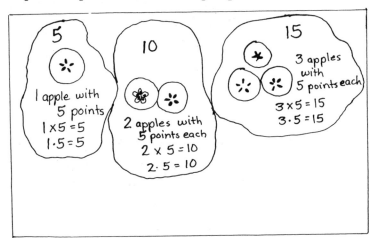

Ask the children to visualize the 10th group in this display. Have volunteers build this group without building those in between and discuss reasons for their answer. Here is one picture of the 10th group.

19 Apples—Extended Number Pattern (continued)

Now have the children determine the total number of points that are in the 10th group and share their thinking.

I see 10 apples with 5 points each. So just add ten 5s (or multiply 5 x 10) to get the total. that's 50 points.

Two apples have 10 points. I see 5 pairs of apples, so I counted 10, 20, 30, 40, 50. There are 50 points.

10 apples with 5 points each
10 x 5 = 50
10 · 5 = 50

This type of imaging may be extended: What does the 20th group look like and how many points does it have? How about the 33rd group? How many apples will be in the group that has 75 points?

Teacher Tips — Encourage the children to look for different ways to total the points in a group. Continue to ask, "Does anyone have another way to share?" or "Can you find another way?"

Journal Writing — Make a list of objects that are naturally seen in sets of five.

Additional Ideas

20 Modeling the 5 Counting Pattern with Rectangles—Part I

You Will Need
- Chapter 3, Patterns; Chapter 6, Multiplication and Division
- colored squares for the overhead and markers
- base ten counting pieces and linear units for each child
- ten 3 × 5 index cards or pieces of paper for every two to four children

Your Lesson

Contact Lesson 19 was the first of six lessons that focus on the counting progression: 5, 10, 15, 20, 25, ... As was true in Lessons 5–10, during which the 2 counting pattern was explored, the goal is to help your children think about this progression in terms of pictures, patterns and song. Today, this is done by forming rectangles with units from the base ten number pieces.

Introduce the 5s counting pattern by reviewing the patterns of appleseeds discussed in Lesson 19. Build the following rectangles at the overhead (illustration A) and have your children work at their desks in groups of two to four to do the same.

Ask them to continue building rectangles of this sort (illustration B). Since these are intended to be a model for the 5 counting pattern, each rectangle should have 5 rows (or 5 columns). Each should also have 5 more units than the one before it.

Discuss the areas of these rectangles with the children as you circulate among them. Pay close attention to the variety of methods used to determine the area of each rectangle. These areas show the counting progression: 5, 10, 15, 20, 25, 30, 35, 40, 45, 50. Be sure to note this connection (illustration C).

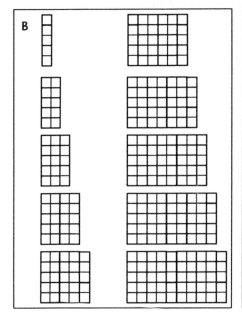
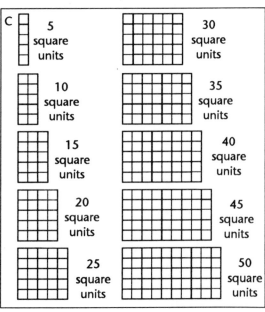

20 Modeling the 5 Counting Pattern...(continued)

Now have the children use linear units to show the dimensions of these rectangles (illustration D). Discuss the patterns of dimensions: one is always 5, while the other increases by 1 unit each time.

Complete the lesson by having the children label the dimensions and area of each rectangle on an index card (illustration E).

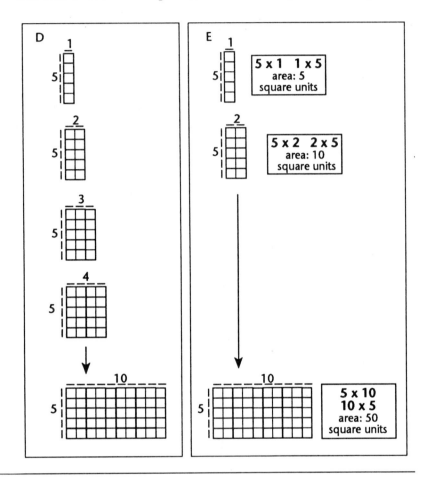

Additional Ideas

21 Modeling the 5 Counting Pattern with Rectangles—Part II

You Will Need
- Chapter 3, Patterns; Chapter 6, Multiplication and Division
- base ten counting piece units
  for each team of two
- several sheets of centimeter graph paper (Blackline 25)
- 18 × 24 construction paper
- glue

Your Lesson

Do the children recall the rectangles built during Lesson 20? To help children bridge the gap between manipulatives and mental images, have them work in pairs to build once more the progression of rectangles using the base ten units. When they are finished, have them prepare a visual display of the rectangles by cutting them from graph paper and gluing them to construction paper. The rectangles should be arranged in area from smallest to largest, as shown below. The dimensions and areas should also be indicated. Save these displays for the next lesson.

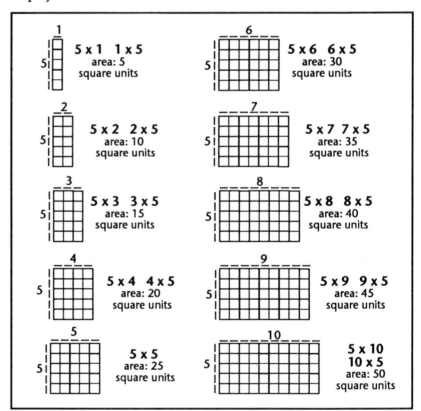

Have the children share their completed displays with one another. Challenge them to predict the areas of other rectangles that would appear in the pattern of 5s.

Teacher Tips

Encourage the children to work cooperatively on this activity and to practice the 5 counting pattern.

22 Further Exploration of the 5 Counting Pattern

You Will Need
- Chapter 3, Patterns; Chapter 6, Multiplication and Division
- the displays of rectangles from Lesson 21
- colored squares for the overhead and a set of linear units *for each child*
- base ten counting piece units,
- large laminated hundreds matrix (Blackline 38)
- individual hundreds matrix (Blackline 27)
- crayons

Your Lesson

Are your children beginning to "see" mental pictures of the rectangles that show the 5 counting pattern? Before beginning today's lesson, practice just a bit.

TEACHER *Close your eyes and try to picture a 4 × 5 rectangle. What is its area?*

CHILD *I see a 2 × 5—that's 10—and 10 more make 20 square units in the area of that rectangle.*

The teacher displays the rectangle at the overhead and invites a volunteer to show its dimensions with linear units.

TEACHER *Thank you. Now try to picture a rectangle that has 40 square units. One of the dimensions is 5. Can you picture this rectangle? Raise your hand when you think you can "see" the other dimension.*

CHILD *I see it! It's twice as big as the 4 × 5! It would have 8 rows of 5 units! The other dimension is 8.*

Return the displays from Lesson 21 to be used as a model and lead your children in more exploration of the 5 counting pattern. Some snapping and clapping would be appropriate here.

TEACHER *I am going to snap and clap the 5 counting pattern. I will snap on all numbers that are not in this pattern, but when I come to one that is, I will clap. Join me when you "feel" the pattern.*

    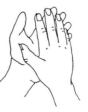

snap, snap, snap, snap, CLAP!

22 Further Exploration of the 5 Counting...(continued)

When the children have found the rhythm to this, ask them to count at the same time, whispering on all snapped numbers and *yelling* on all clapped ones. Discuss the fact that the claps occur on every 5th number (or multiple of 5) and that these numbers correspond to the areas of their rectangles.

Challenge your children to use base ten units to cover the numbers in the 5 counting pattern on a large hundreds matrix (Blackline 38).

1	2	3	4		6	7	8	9	
11	12	13	14		16	17	18	19	
21	22	23	24		26	27	28	29	
31	32	33	34		36	37	38	39	
41	42	43	44		46	47	48	49	
51	52	53	54		56	57	58	59	
61	62	63	64		66	67	68	69	
71	72	73	74		76	77	78	79	
81	82	83	84		86	87	88	89	
91	92	93	94		96	97	98	99	1

What patterns do they notice in the arrangement of squares on the matrix? If the matrix was extended, would 130 be covered? How about 403 or 2,000? Why or why not?

Complete the lesson by having each child shade in the 5 pattern on a small hundreds matrix to keep as a permanent record. These can be glued to their displays of rectangles for this counting pattern.

Teacher Tips

You may choose to keep the finished displays at school for a while. However, don't be surprised if your children insist on taking them home to be shared by all!

Compare the 2 and 5 counting patterns. Challenge your children to collaborate in determining likenesses and differences between these patterns. They may make some interesting discoveries!

Additional Ideas

23 5s—Dimensions and Areas; Song

You Will Need
- Chapter 3, Patterns; Chapter 6, Multiplication and Division
- A classroom set of Multiplication/Division Discussion Cards
- number pattern song cassette tape, *Musical Array-ngements*

Your Lesson

As you did at the start of Lesson 22, provide some time for children to mentally picture rectangles that have a dimension of 5. Can they "see" these rectangles and identify their dimensions and areas? Ask questions such as:

Can you picture a 5 × 4 rectangle? What is its area?

Is it possible to arrange 50 square units into a rectangle that has 5 rows? If so, what would such a rectangle look like? If not, why not?

Can you give an example of a number of units that could not be arranged into a rectangle that has 5 rows?

What are some large numbers that will be part of the 5 counting pattern?

Can your children identify rectangles having a dimension of 5? Try it! Randomly distribute a set of Multiplication/Division Discussion Cards, giving each child several cards. Invite anyone holding a rectangle that has a dimension of 5 to come forward. When all such rectangles have been located, have the owners of the cards arrange themselves so that the cards are ordered from smallest area to largest.

Can each of these children name their rectangle by its dimensions? Allow for a moment of discussion (so that no one feels put on the spot), and then ask the children to "sound off": "1 by 5", "2 by 5", "3 by 5", etc.

23 5s—Dimensions and Areas; Song (continued)

Can they state the areas of their rectangles? Again, allow some time for collaboration, then: "5", "10", "15", "20", etc.

Encourage the entire class to join in the pattern. Everyone can clap, snap or chant as the card holders raise their cards proudly at the appropriate time.

Let the good times roll! Play the song, "Slicing by Fives", from the number pattern cassette tape. Hope your children love spittin' those seeds as much as ours do!

Sing the song several times, trading cards after each sequence. Also, accept requests for "Two's Shoes." Our children love to review the two songs and compare for common numbers. Which numbers appear in both songs? Why?

Invite your children to make some predictions about the cards. If the pattern of cards were continued, what would the next rectangle look like? What would be its area? How about the rectangle on the 30th card—can someone describe its dimensions and area? How about the rectangle on the 100th card? Etc.

Teacher Tips Have the children sing the songs with their eyes closed. Suggest they form mental images of the rectangular arrays that correspond to the numbers being sung.

Additional Ideas

24 Mystery Box—Football; 5s—Using a Calculator

You Will Need
- Chapter 3, Pattern; Chapter 6, Multiplication and Division
- your mystery box with a football in it
- a calculator for each child
- number pattern song cassette tape, *Musical Array-ngements*

Your Lesson

Kick off the football theme (which begins in Lesson 30) sometime during this day (not necessarily during math time). Challenge your children to guess the contents of the Mystery Box by following the procedures described in Lesson 1. You might vary the manner of keeping track of the questions. Perhaps this month a stack of 20 hex-a-link cubes could be used. As each question is asked, a cube is removed from the stack; when all are gone, questioning ceases until the next day.

5	10
15	20
25	30
35	40
45	50

Sing "Slicing by Fives" with your children and write the 5 counting pattern on the chalkboard. Is this song beginning to rival "Two's Shoes" in popularity?

Ask the children to extend this pattern, using calculators if necessary. Have them record the numbers and examine them for patterns. Discuss questions such as the following:

What patterns can be observed? Do they notice, for example, the 5–0 repetition that occurs in the units' digit?

How can one predict if a number will appear in the 5 counting pattern? What about numbers like 124? 450? 4,567? etc.

How many 5s make 60? Make a prediction and check it with a calculator, perhaps by using repeated addition. Draw a rectangle on grid paper that shows this answer.

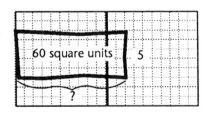

Do the same for other numbers. How many 5s make 90? 135? etc.

Imagine (or sketch) a rectangle that has 5 rows of 24 square units. How many square units are there altogether?

Imagine (or sketch) a rectangle that contains 85 square units and that has 5 rows. What is the other dimension? How can a calculator help here?

24 Mystery Box—Football...(continued)

Be sure to allow time for the children to share their observations and thinking about these questions.

End by singing "Slicing by Fives" once more.

Teacher Tips

Ask the children to discuss the ways they were able to extend the 5 counting pattern. Did the song help? Did any think of pictures of rectangles that have a dimension of 5? Did any feel the need for a calculator?

Your children may want to record the 5 counting pattern as the numbers appear on a calculator. If so, encourage them to loop any patterns of repeating digits that are observed. This can help them check for accuracy.

```
00    50    100
05    55    105
10    60    110
15    65
20    70
25    75
30    80
35    85
40    90
45    95
                Mike
```

Additional Ideas

25 Geometry—Wind Socks!

You Will Need
- Chapter 3, Patterns; Chapter 11, Geometry
- 3 or 4 rolls of crepe paper streamers
- paper scraps, markers or paint for decoration

for each child
- a sheet of 12 × 18 construction paper
- glue and scissors

Your Lesson

Today, the children will create wind socks to brighten up the room! Ask them to form a cylinder by bending a piece of construction paper and then gluing the sides.

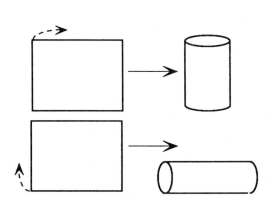
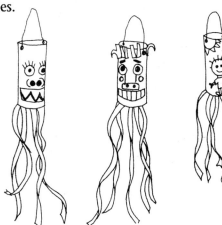

Model a rich mathematics vocabulary (as always) when teaching this lesson. Use words such as cylinder, hollow, parallel, etc., in the discussion, so these words take on specific meaning.

Have the children demonstrate their procedures and then share observations about their cylinders. They might comment about the circular base or that they are reminded of cans or pipes.

Discuss the fact that the paper cylinders are hollow, as are wind socks. Have the children ever seen wind socks? What purpose do they serve? Why are they hollow? Why do they have streamers attached to them?

Are wind socks ever solid? Why or why not? Why are they always long and skinny, rather than short and fat?

Ask the children to add streamers to their cylinders to create a wind sock. They will enjoy decorating them in various ways.

Teacher Tips

This activity may serve as both math and art time. The windsocks add cheer to many days while they are on display.

Additional Ideas

26, 27 Geometry—Symmetry

You Will Need
- Chapter 11, Geometry
- overhead pattern blocks
- chart paper and markers
- a container of pattern blocks available for every 2–4 children

Your Lesson

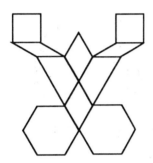

Children are generally fascinated with pattern blocks and enjoy creating designs with them. Informal activities with these blocks can strengthen an awareness of shapes and their properties. In this two-day lesson, children use the blocks to learn about symmetry.

You may wish to begin by letting the children explore freely with the blocks. Then, after an appropriate time, display the design that is pictured to the left.

Invite children to share things they notice about this arrangement. Be sure to discuss the symmetry that is present, and use this design to motivate an exploration of this concept (please refer to the dialogue in Lesson 1 of Chapter 11, Geometry, pages 86–88).

Continue the lesson by dividing the class into small groups and asking each group to create other pattern block designs that have symmetry. Can they make some that have one line of symmetry? Two? Three?

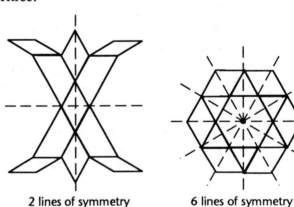

2 lines of symmetry   6 lines of symmetry

Journal Writing

Have the children comment about their designs or write summaries of the work. For example, how would they describe the location of the pieces in relation to the line of symmetry? Do they recognize that a line of symmetry separates the design into two parts that are mirror images of ea

Copycat Game

26, 27 Geometry—Symmetry (continued)

Teacher Tips

Here is another group activity: Have one member of the group make a design and identify a line that is to be a line of symmetry. The others in the group can then play "copy cat" by working to create the full design!

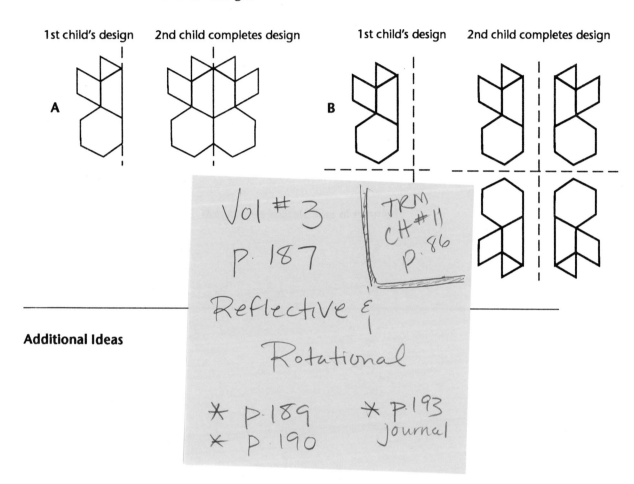

Additional Ideas

(handwritten note:)
Vol #3
p. 187
TRM CH #11 p. 86
Reflective &
Rotational
* p. 189   * p. 193 Journal
* p. 190

28 Circumference—Estimation and Comparison

You Will Need
- Chapter 7, Measurement
- a large circle and a length of string that matches its circumference
- for each group of 4 or 5 children: 6 circles and lengths of string that match their circumferences

Your Lesson

This lesson is intended to strengthen children's intuition about circumferences of circles.

Hold up a piece of string and tell the children that it will fit exactly around a circle that you are hiding. Ask them to imagine this circle and to draw a sketch of it on their paper. How big do they think the circle will be? Now show the circle and wrap the string around it. How close were the children's guesses?

Continue by giving each group of four or five its set of 6 circles and strings. (Before distributing these, label each circle and string with a letter or number to aid in recording guesses.) Have the groups try to match the strings with the corresponding circles. Encourage them to make predictions before wrapping the strings around the circles.

Journal Writing

Have the children write observations about their work in their journals. They might notice, for example, that shorter strings match with smaller circles or be reminded of other times when they wrapped strings around circular objects. Ask them, too, to examine some of the other circles present in the room. Do they think longer or shorter strings would be needed to wrap around the top of the wastebasket or around the clock?

Additional Ideas

29 Volume

You Will Need
- Chapter 7, Measurement
- 8–10 containers, including some that appear to have the same volume
- suitable material for filling the containers (rice, water, sand)
- measuring cups (optional)

Your Lesson

Show the containers to the class. Brainstorm ways to measure the volume. Would sand, rice or water be helpful? How about base ten pieces or wooden cubes? Why or why not?

Which do they predict will hold more liquid (or dry) measure? We like to use rice for measuring volume; however, water is cheaper, more available and more fun, if you're into messes!

TEACHER *I have here a pickle jar and a peanut butter jar. Which do you estimate has the greater volume?*

CHILDREN *The peanut butter jar!*

TEACHER *Let's put the container we predict has the greater volume on this end and the smaller volume on this end. Now, examine this vase that held flowers from Austin. Where shall we put it in the order?*

Place it first on the low end and let children raise their hands to vote if they feel this placement is appropriate. Move it to the middle. Vote again. Move it to the end reserved for those with the greatest volume. Vote again. Place it where most students feel it belongs.

Discuss the placement of each container in the same manner.

29 Volume (continued)

Brainstorm ways to test the agreed upon order. For example, some of our children suggested filling each with rice and pouring the contents into a large clear container. They then suggested marking the side of the container to indicate the level of rice. "How handy!" someone reasoned, "Measuring cups come with a 'ruler' printed on the side! Awesome!"

Another strategy is to compare the jars in pairs by pouring rice from one into the other.

Hey! I still have some rice left over, so the square round container must hold more than the peanut butter jar.

How accurate were the children's predictions? Did any container(s) stump them? If so, what caused the difficulty?

Homework

Have your children bring containers from home to try and stump their friends tomorrow. This is an easy way to gather a large variety of containers for comparison.

Additional Ideas

30 Volume

You Will Need
- Chapter 7, Measurement; Chapter 10, Data Analysis and Graphing
- various containers children have brought from home
- appropriate material for filling the containers
- a copy of Blackline 81 (Observation Record Sheet) for each child (optional)

Your Lesson

This lesson extends the volume activity in Lesson 29. Ask each child to compare the volume of their container with that of a friend. Have them first estimate which holds more and then make a test using filler material. Each should keep a record of the result, perhaps with a graph similar to the one below. As time permits, the children can rotate to make similar comparisons with others in the class.

Container Comparisons

less capacity than my container	✓	✓	✓	✓	✓	✓	
equal capacity	✓	✓	✓				
more capacity than my container	✓	✓	✓	✓	✓	✓	✓

Journal Writing

As usual, allow enough time for an analysis of the data. Each child can write a few sentences about the information shown on the graph (above) in their Math Journals. They might comment about the nature of their results on their Observation Record Sheet, noting the accuracy of their predictions and if they found any of the containers deceiving.

Additional Ideas

31 People Fraction Bars

You Will Need
- Chapter 8, Fractions
- a set of People Fraction Bars, including a whole uncut bar of each color. (See the Materials Guide.)

Your Lesson

People Fraction Bars are useful for helping children relate fractions to information about themselves. In the following activities, children think about the meaning of fractions and form pictures of them with the bars.

Discuss the construction of People Fraction Bars with your class. Have volunteers demonstrate that the uncut green, yellow, blue, purple and red bars are the same size. Tell the children that another green bar of the same size has been cut into 2 equal parts. Hold up these parts and show how they can be placed side by side to exactly match an uncut bar; describe the parts in the same way the yellow, blue, purple and red bars have been cut into.

Give the children an opportunity to compare the lengths of the various parts that have been cut and discuss questions such as:

Why are the parts cut from a green bar bigger than the ones cut from a yellow bar? (The green and yellow bars are the same size, but the green bar has been cut into fewer equal parts.)

Which part is the smallest? Why? (The red part is smallest because the red bar has been cut into more equal parts than any of the other bars.)

How are the yellow parts related to the red ones? (The yellow parts are twice as big as the red ones. This is because the red bar has been cut into twice as many equal parts.)

It is important to note that the uncut bars are the same size. Because of this, the more parts a bar is cut into, the smaller each part will be.

It is now time to use the bars to form fractions related to the children. The following dialogue describes how this can be done. Four children come to the front of the room, each child is holding 1 of the 4 parts cut from a blue bar. They show the unshaded sides to the class.

31 People Fraction Bars (continued)

TEACHER *What do you notice about these children?*

JOSH *Well, 3 are girls.*

The 3 girls turn their bars over so that the slashed sides are exposed. The 4 children push their bars together as shown here.

The class discusses such observations as:

Three out of the 4 children, or ¾ of the children, are girls.

The 4 parts of the bar represent the 4 children in the group.

The 3 slashed parts that are exposed represent the 3 girls in the group.

MARIE *I see something else. None of them are wearing sweaters!*

All 4 children show the unslashed sides once more.

Other information about the group can be similarly portrayed and discussed.

Two of the 4 children, or ²/4 of the children, have glasses.

All of the children are standing, so ⁴/4 of the children are standing. This bar is called a "whole" bar.

Repeat the activity using groups of 2, 3, 5 or 6 children. Note that the number of children in the group determines which color of Fraction Bar is used, i.e., a group of 6 children will use red bar parts to form its fractions.

Include examples that lead the groups to form zero bars and whole bars, as shown above.

31 People Fraction Bars (continued)

Emphasis

Describe each fraction with appropriate language, identifying the whole to which it refers. For example, "Each yellow part is $1/3$ as big as a whole (uncut) bar." and "Five-sixths of the children ate cereal this morning."

The denominator in each people fraction refers to the number of children in the group being described. The numerator identifies how many of those children satisfy a given criterion.

When no child in the group meets the specified criterion, a zero bar is formed. When everyone meets the criterion, a whole bar is displayed.

The more equal parts into which a bar is divided, the smaller each part will be.

Teacher Tips

Consider using these bars when forming people fractions during your Calendar time.

If purple posterboard is not available, you might mark white posterboard with heavy purple lines.

Additional Ideas

32 Diagramming People Fraction Bars

You Will Need
- Chapter 8, Fractions
- a set of People Fraction Bars
- copies of Blackline 1 (Friendly Fractions) for each child
- an overhead transparency of Blackline 1

Your Lesson

Divide your children into groups of three. Ask one of the groups to use the yellow parts of a People Fraction Bar to show a fraction about its members. Display the transparency of Blackline 1 and tell the children that each rectangle represents an uncut People Fraction Bar. Divide the first rectangle into 3 equal parts and diagram the fraction that has been formed (an example is shown here).

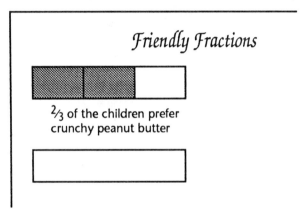

Ask the children to make a similar diagram on their blacklines and discuss it. What do the 3 equal parts represent? What do the slashed parts portray?

Do the same for a few other People Fractions using different groups of 3 children.

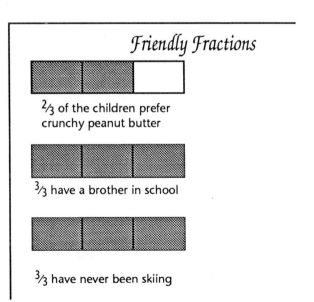

32 Diagramming People Fraction Bars (continued)

Have a group of 4 children form a People Fraction and ask the children at their desks to diagram it. Invite a volunteer to describe their diagram at the overhead.

Repeat this activity using other groups of (2, 3, 4 or 6) children.

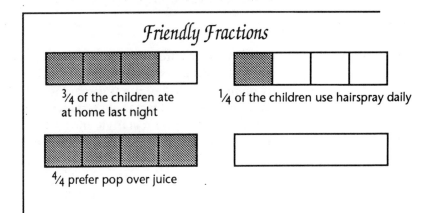

Emphasis

As each fraction is diagrammed, identify what is represented by the numerator and denominator.

On Blackline 1, each rectangle represents an uncut bar. It is then divided into equal parts that correspond to the parts of a People Fraction Bar.

Additional Ideas

33 Introduction to Group Fraction Bars

You Will Need
- Chapter 8, Fractions
- a set of People Fraction Bars *for each group of four children*
- a set of Group Fraction Bars (see Materials Guide)
- copies of Blackline 2 (Open-Ended Fraction Bars)

Your Lesson

Group Fraction Bars are similar to People Fraction Bars, only they are smaller and are cut from construction paper. Children use them to form fractions about themselves while working in small groups.

One way to introduce Group Fraction Bars is to divide your class into groups of four and invite one of the groups to the front of the room to form a People Fraction suggested by the class.

Now have the other groups show the same fraction at their desks, using the blue parts of their Group Fraction Bars. When this has been done, ask each child to diagram the fraction on Blackline 2.

2 of the 4 of us have brothers attending this school.

Repeat this activity, each time asking a different group of children to form a people fraction in front of the class. Note that when a group of 3 creates the fraction, then the children who are seated will have to use the yellow parts of their Group Fraction Bars.

As time permits, ask groups of two to six children to form the People Fractions so that the green, purple and red parts of the Group Fraction Bars are also used.

Additional Ideas

34, 35 Group Fraction Bars

You Will Need
- Chapter 8, Fractions
- Group Fraction Bars and copies of Blackline 2 (Open-Ended Fraction Bars) for each small group of children
- transparency of Blackline 2
- feely box
- a copy of Blackline 60 (Coin Toss) for each child for homework preliminary to Lesson 39.

Your Lesson

Day 1

Divide the class into small groups of three, four, five or six children. Ask each group to form fractions about itself with its Group Fraction Bars and to make diagrams of these fractions on their blacklines.

Be sure to allow time for the groups to share some of their diagrams at the overhead.

Conclude the day's activities by asking each child to write on a slip of paper a criterion for forming fractions. Put these papers in a feely box to be used in tomorrow's lesson.

Day 2

Divide the class into groups of three, four, five and six children. Draw a slip of paper from the feely box containing the suggestions for forming fractions that the children wrote yesterday. Ask the groups to form fractions with their Group Fraction Bars using the criterion that is on the slip of paper. When ready, have the group members diagram their fraction on Blackline 2.

As time permits, repeat this activity using other suggestions from the feely box.

34, 35 Group Fraction Bars (continued)

Emphasis

In these activities, it is important that the groups decide which Group Fraction Bar is appropriate for each task and that they discuss their diagrams. This helps children clarify their understanding of the role played by the whole group of children (and the whole Fraction Bar) and how their fractions are related to this whole.

Teacher Tips

As time permits, repeat this activity, rotating the children so that they have opportunities to work in groups of three, four, five and six.

Our children enjoyed describing criteria for forming fractions. Their ideas included "Your father's middle name begins with R", "You have visited our state capitol", and "You know how to play Crazy Eights".

Homework

In anticipation of the probability activity that will be conducted in Lesson 39, give your children a copy of Blackline 60 (Coin Toss) and ask them to perform the following experiment tonight and the next three nights: Each night, flip a coin once for each member of the class. On the blackline, keep track of the number of heads and tails that occur.

Journal Writing

Each child can prepare a brief report about the fractions formed within their group. Encourage the use of diagrams in these reports.

Additional Ideas

36 Football—Geometry

You Will Need
- Chapter 11, Geometry
- a transparency of Blackline 59 (Football Field)
- at least 3 copies Blackline 79 (Football Field Recording Sheet) for each pair of children

Your Lesson

How many rectangles can be found on a football field?

Begin an exploration of this question by displaying a transparency of a football field and inviting volunteers to indicate some of the rectangles they see. Students can record these answers on Blackline 79.

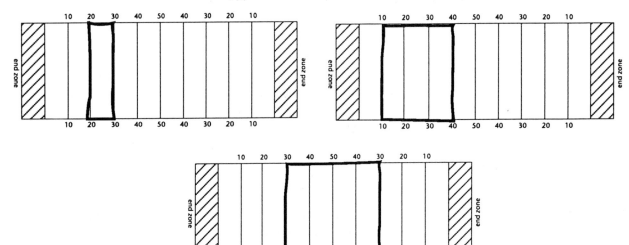

Now have students work in pairs as they go on a rectangle hunt. How many rectangles can each team find? How can a count of the rectangles be systematically made? Teams should shade or outline the rectangles that are found on Blackline 79. At the end, have teams share their counting strategies with the class.

Teacher Tips

Some teams may generate rectangles in a random way. Others may find ways to organize their work. One strategy is to count rectangles in order, according to their dimensions. That is, first tally those 10 yards wide; then those 20 yards wide; etc.

Have your local band director speak to your class about band formations on the football field. How do they use shapes to plan a half-time performance?

Note: For Lesson 37, you may want to adjust the lesson and ask the children to bring in a variety of footballs (small plastic ones given away as promotions, sponge footballs, pee-wee, regulation size, etc.) or other balls (tennis, soccer, golf, etc.).

Also, remind your children to continue gathering data for the coin-tossing experiment announced yesterday.

37 Football—Measurement

You Will Need
- Chapter 7, Measurement
- a football and string
- graph labeled "Too Long, About Right, Too Short" (see below)
- scissors and paper for every child

Your Lesson

Thought we were going to have you measure a football field, eh? Why, no, just measure the football itself.

Hold up the ball that finally escaped from your Mystery Box. Tell the children that you are going to wrap a piece of string around the belly of the ball as shown here.

Ask the children to estimate how long the string should be in order to wrap exactly once around the ball. They might find it helpful to think in terms of hand spans as they do this. How long is their span and how many spans do they think will go around the ball?

Have each child cut a strip of paper that has a length equal to their estimate and to write their name on it.

Wrap the string around the ball and ask a volunteer to cut a piece of paper that is as long as the required length. Let each child compare their paper estimate with the actual length and include it on a class display as shown here:

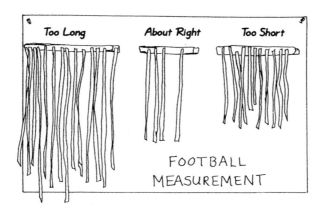

37 Football—Measurement (continued)

What observations do the children have about this graph? Also, discuss the entire experiment. Did the children experience any difficulties while estimating? Do they have any suggestions for overcoming them? For example, they might find it helpful to have the required length marked on the string and then have the string laid out straight prior to making their estimates.

Teacher Tips

Does it matter which way a football is measured? Will the distance around the ball always be the same? A nice extension to this lesson is to have the children decide on the longest path around the ball.

This lesson can be varied by having your children estimate and measure the distance around footballs (or other types of balls) of different sizes. Perhaps they can arrange the balls in order from the smallest circumference to the largest. They can also graph and write about their results.

Additional Ideas

38 Football—Money

You Will Need
- different catalogs that advertise footballs
- money feely boxes, record sheets and calculators

Your Lesson

Has football mania taken over your classroom yet? Perhaps your children will want to buy one to donate to the school!

In groups of four, have the children explore the manner in which department store catalogs are organized. In particular, have each group search for a football that suits its needs, lay out and record different combinations of money from their feely boxes that could be used to purchase it.

Teacher Tips

We like to have each group report to the class about its work. The report usually includes a description of the ball that has been selected, some of the money combinations that could be used to pay for the ball, and some related story problems.

Reminder: Be sure to remind your children to bring their data from the coin-tossing homework experiment to school tomorrow.

Additional Ideas

39 Football—Statistics

You Will Need
- Chapter 9, Probability; Chapter 10, Data Analysis and Graphing
- materials for graphing
- a copy of Blackline 81 (Observation Record Sheet) for each child

Your Lesson

Over the past four nights at home, your children have been gathering data for the following experiment: Flip a coin once for each member of the class and tally on Blackline 81 the number of heads and tails that occurred.

When the children return to school today with their completed record sheets, divide them into groups of four and have them make written observations about their data on their record sheets. Some of the questions that might be addressed during this time are:

Which occurred most often for each child, heads or tails?

How does the data obtained by the different group members compare? Are there any similarities or differences? How might the findings be explained? What appears to be a "representative" number of times that heads occurred with the group?

Have each group prepare a graph that shows the combined results of its members. How do these results compare with those of each person?

If the activity were repeated 1,000 times, how many heads might be expected?

The collective results of each group may be compared in a similar manner. In addition, determine the total number of times heads or tails occurred for the entire class. Did either occur more frequently? Did the same thing happen in the individual or group results? Why or why not? What might explain why both teams in a football game regard the opening toss of a coin as fair?

Allow time for each group to complete their observation record sheets.

39 Football—Statistics (continued)

Teacher Tips

Theoretically, in the long run, heads should occur as often as tails if a fair coin is flipped in an unbiased manner. This should be especially noticeable in the combined results of the entire class. It is possible, however, that individual (or group) results in this experiment may seem to favor one outcome over the other. When this happens, it might be due to weighted coins (or biased tossing) or to having a sample of results that is too small.

Deciding on a "representative" number of times heads occurred within each group can be an intuitive way to think about an average. You may wish to incorporate this activity into the lessons on averages within this program and have the children determine the average number of times heads occurred in each group or in the entire class.

Additional Ideas

40 Football—Problem Solving

You Will Need
- a local newspaper containing football scores of interest to the class
- a chart that shows the following ways to score points in a professional football game

> Touchdown: 6 points
> Conversion after a touchdown: 1 point
> Field Goal: 3 points
> Safety: 2 points

Your Lesson

After a brief discussion of the results of recent football games, ask the children to suggest different ways of answering the question: How can a professional team score a total of 17 points?

Do the same for other scores. In a professional game, is it possible for a team to score 19 points? If so, in what way(s). How about 56 points? 199 points? Is there any point total that cannot be achieved in some way?

It might be helpful to record answers as shown here:

		Points			Points			Points
6	Touchdowns	36	__	Touchdowns	__	5	Touchdowns	30
6	Conversions	6	__	Conversions	__	5	Conversions	5
4	Field Goals	12	__	Field Goals	__	7	Field Goals	21
1	Safeties	2	28	Safeties	56	__	Safeties	__
	Total:	56		Total:	56		Total:	56

40 Football—Problem Solving (continued)

Theoretically, every point total except 1 can be achieved and most can be made in many ways. Children generally enjoy the challenge of finding ways to score large number of points, pretending that a game can go on and on. They often think about these questions in different ways and these should be shared with the class.

Teacher Tips

Several calculating options may be highlighted in this lesson. For example, smaller combinations of points may be mentally computed or modeled with number pieces. Larger combinations may also be determined mentally or with a calculator.

Homework

(Optional) Select a score from a game of interest in your area. Have your children brainstorm at home several ways that such a total could have been obtained.

Additional Ideas

41, 42 Football—Story Problems

Your Lesson

One of our children's favorite activities is to make a book of story problems about football. One way to do this is to have each child glue a favorite football picture on an 11 × 18 piece of paper and then write a problem related to the picture. The solution to the problem is covered by a flap and then the problems are assembled into a book.

We usually wait until the next day to read the stories. At that time, the children try to solve each problem, comparing their solutions with the author's. Very often solutions will be reported in different ways and lots of problem solving (and giggles) will take place.

Teacher Tips

Some children may prefer to write a story first and create a picture to accompany it. These may be about popcorn, cheerleaders, bands or other things related to football.

Encourage those children who especially enjoyed Lesson 40 to create a similar story problem involving the way(s) a score was obtained.

Additional Ideas

43 Football—Fractions/Measurement

You Will Need
- Chapter 8, Fractions
  for each child (or group)
- 60 hex-a-links and a paper circle
- Group Fraction Bars

Your Lesson

How long is a football game? How is the time divided up?

Have the children model the periods of a game in different ways. Our children chose hex-a-links, circles and fraction bars for this purpose. Each child (or group of children) is asked to make a stack of 48 hex-a-links and to place it, a circle, and a completely shaded fraction bar on their desk. These represent the time needed to complete a high school football game (in Kentucky, this is 48 minutes).

The children then seek to represent the halves and quarters of the game. For example, a half might be modeled by half of the circle or by a fraction bar that is only half shaded.

Repeat this activity for college/professional games that span 60 minutes or for the pee-wee football games in your area.

As time permits, other questions can be considered. For example: If a game lasts 80 minutes, how long is each quarter? In hockey there are 3 periods. How long is each period if the game lasts 60 minutes? How can these periods be modeled with the materials used in this lesson? A professional soccer game lasts 90 minutes and is played in 2 halves. How long is each half?

43 Football—Fractions/Measurement (continued)

Teacher Tips

Get a television schedule that tells the amount of air time that is allotted for a college game. Why is that much time needed?

You might elect to incorporate weekly football scores into the calendar activities for a few weeks. Have your children select a team of interest and prepare a graph that shows its record at the end of each week. Newspaper clippings and other observations about the team can be posted beside this graph.

Additional Ideas

44 Football—Problem Solving and Patterns

You Will Need
- base ten counting pieces, grid paper and calculators (as needed)

Your Lesson

Imagine a football extravaganza in your home town called the Field Goal Bowl. It is played like a regular game of football except only 3-point field goals are allowed; this provides a motivating context for working with multiples of 3. Here are some problems that might be posed.

What scoring possibilities would there be in such a game? Could a team ever score 15 points? How about 29 points? How many points would a team score if it made 14 field goals? 38 field goals? How many field goals would it take to score 66 points? 69 points? 261 points? Give some examples of other point totals that could not be made.

Challenge the children to explore different ways for solving these problems and invite them to share their thinking. Encourage them to model some of the problems with counting pieces or sketches. Answers may also be obtained by using mental arithmetic or calculators, whenever appropriate.

44 Football—Problem Solving and Patterns (continued)

Teacher Tips

You may wish to begin this lesson by having the students review their knowledge of multiples of 3. One way to do this is to have each child write, "Field goal! Score 3 points for me!" on a football shape. These can then be displayed in a manner that highlights multiplication by 3.

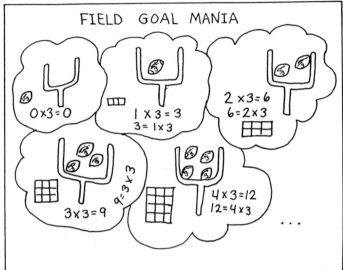

There are other pattern activities that could be used with the theme of football. One such activity is to have your children bring in pictures of football action and create patterns with them. For example, they could create an AABAC pattern such as, "helmet on, helmet on, helmet in hand, helmet on, helmet not pictured."

Additional Ideas

45–49 Exploring the 3 Counting Pattern

Your Lesson

The activities of yesterday's Field Goal Bowl (Lesson 44) can be used to motivate an exploration of the 3 counting pattern: 3, 6, 9, 12, 15, 18, ... This exploration may be conducted in the same manner as those for the 2 and 5 counting progressions (Contact Lessons 6–10 and 20–24, respectively). General guidelines for these lessons are also detailed in the booklet, *Opening Eyes to Mathematics Through Musical Array-ngements*.

Teacher Tips

The first part of this exploration consists of modeling the 3 counting progression with rectangular arrays that have a dimension of 3. It is helpful to connect this activity with those of Lesson 44, perhaps with questions such as: Which array could be used to picture the total number of points scored by 7 field goals? How many field goals must be kicked in order for a team to score 30 points? Which array pictures this? How many are needed to score 60 points? Etc.

There are many patterns to be noticed when the multiples of 3 are shaded on a hundreds matrix. Here are some observations made in our classroom:

"When I look at the rows, I see a skip-skip-cover pattern. I also see it in the columns, too."

"I notice when I count by tens, 30, 60 and 90 are covered. That's every third one. One hundred won't be covered; the next one will be 120."

"I know 99 will be covered because 99 can be divided by three 33 times!"

Complete the exploration of the 3 counting pattern with some calculator activities. For example, you might ask such questions as: how many 3s make 60; make a prediction and check it with a calculator, perhaps by using repeated addition; draw a rectangle on grid paper that shows this answer; do the same for other numbers; how many 3s make 93? 135? etc.

Imagine (or sketch) a rectangle that has 3 rows of 24 square units. How many square units are there altogether?

Imagine (or sketch) a rectangle that contains 105 square units and has 3 rows. What is the other dimension? How can a calculator help here?

50 Mystery Box—Money; Geometry—Body Shapes

You Will Need
- Chapter 11, Geometry
- Mystery Box with some coins in it
- hexagons and triangles from a set of pattern blocks
- the bodies of you and your children!

Your Lesson

Money-money-money! Don't we all have those days when we wish for pennies from heaven? Although we might prefer that, finding a large collection of pennies (or mixed coins) inside your mystery box might serve as a welcome substitute. The mathematics to follow is certain to be rich in rewards!

Sometime during this day begin the mystery box process to introduce the theme of money (which starts with Lesson 57). We establish a quota of 18 questions per day. We place 3 hexagons on the overhead and tell the children that each question will cost 1 triangle (this will require some trading of the pieces). When the hexagons have all been removed, the box is put away until the next day.

Now to get rid of some stress by stretching muscles, laughing and creating body shapes! Have your children spread out on the floor and let it happen! Have them form triangles, rectangles, circles, cylinders, parallelograms, etc. Here are possibilities:

A triangle using the floor as a base and leg bent at the knee for the other sides.

A circle formed with a mouth or by having everyone join hands (or sit) appropriately.

Rectangles or parallelograms using hands or feet.

A right angle created by sitting erect on the floor.

50 Mystery Box—Money; Geometry (continued)

Tell the children to grab a partner. Can the two of them form a rectangle using only right arms; a rectangle by joining feet; a parallelogram by joining feet; a circle by joining fingertips; a cylinder (not quite solid) by joining fingertips and feet; a triangle with their backs as two sides and the floor as the third side?

The possibilities are endless. Let them come up with creations of their own and challenge classmates. Giggles are unavoidable as human pretzels appear during this lesson!

Additional Ideas

51 Geometry—Elastic Shapes

You Will Need
- Chapter 11, Geometry
- 2 yards of 1" black elastic sewn together at the ends, one for every three–four children
- chart paper
- markers

Your Lesson

Can your children create shapes from elastic loops?

This lesson is loved by children and adults as well. We borrowed it from Donna, Allyn and Paula's *Box It or Bag It Mathematics* for earlier grades because we found our children love it so. It lets the creative juices flow!

Give a piece of elastic to every three or four children. (We like black elastic because it doesn't show dirt.) Have the children work together to form a square with the elastic. Each group member must be involved in some way. Most groups will hold it with hands at each corner.

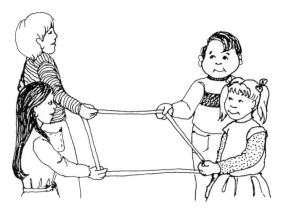

Ask the children to describe an attribute of their square, as you record their responses on chart paper. If they have trouble with this, ask questions such as, "What kind of angles are there?" or "What else do you see?" Here are some of the observations we have heard: "It has 4 sides." "It has 4 angles." "It has 4 right angles." "I see parallel sides." "All sides are the same length." "It's a rectangle and a parallelogram."

51 Geometry—Elastic Shapes (continued)

Challenge them further with exercises such as:

Form a square without using any hands. For example, some may replace their hands with their necks.

How can they make their squares rotate?

What are some ways that their square can be made a different size? (Please see Lesson 2 on pages 88–90 of the Geometry Chapter for some possibilities.)

How can their square be changed into a triangle? You'll know immediately when a child has thought of an answer, because they will surprise the others by dropping their corner!

Teacher Tips   As time permits, repeat this activity by making other shapes such as triangles or pentagons.

Journal Writing   Ask your children to describe in their journals how they formed their shapes. This is a wonderful opportunity for them to do some writing and to improve their mathematics vocabulary!

Additional Ideas

52, 53 Geometry—Shapes

You Will Need
- Chapter 11, Geometry
- pattern blocks available for every three or four children
- overhead pattern blocks
- chart paper

Your Lesson

(This is a two-day lesson. At the end of the second day, you will need molds for creating solid shapes. These will be used in Lesson 54 and are described in the Teacher Tips at the end of this lesson.)

Day 1

Distribute the pattern blocks and ask the children to examine a green equilateral triangle. What do they notice about their triangles? They are likely to make observations such as those shown to the left.

All the sides are equal. There are three equal angles.

Two of these triangles will cover the blue diamond.

Six triangles can form a hexagon.

Make a record of their observations on chart paper (see illustration below) to be used later for reference purposes.

In groups, have the children examine the other pattern blocks in the same way, compiling a list of things they notice about each block. Encourage them to also search for relationships that exist among the different blocks.

3 equal sides
3 equal angles
6 can form a hexagon
It is called an
   equilateral triangle

4 equal sides
Opposite sides are parallel
2 triangles will cover it
It is a parallelogram
Opposite angles are equal

Note: This is a good opportunity to review (or introduce) terms such as parallel, perpendicular, parallelogram, etc. Also, should the children not mention properties that you wish to discuss, point them out yourself. You might say, "Here is something I noticed about the blue diamond: It has 2 pairs of parallel sides. It is an example of a parallelogram." (The blue diamond is also an example of a rhombus, since all of its sides have the same length.)

52, 53 Geometry—Shapes (continued)

Day 2

Ask the children to investigate the following questions:

What shapes can be made with 2 green triangles? What could be made with 3 triangles? 4? 5? etc.

Number of Triangles

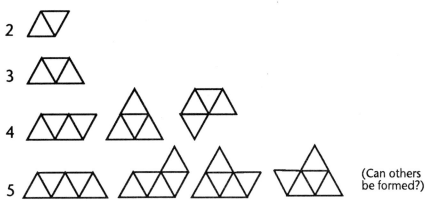

(Can others be formed?)

As time permits, similar investigations can take place using other pieces or combinations of pieces. The act of moving pieces around to create different shapes helps children see how shapes are related and strengthens their spatial awareness.

Teacher Tips

In preparation for Lesson 54, Solid Shapes, you will need to allow some time at the end of the second day of this lesson to have the children create molds that are shaped like a rectangular prism, a cube, a cylinder and a cone. The first two can be cut from milk cartons, the third can be an empty orange juice can and the fourth a paper water cup or sugar cone wrapper. Fill the molds with water and allow them to freeze overnight.

The molds can be filled with substances such as plaster of paris or gelatin instead of water. If you are ready for such an adventure, you may wish to take time from the day's Insight lesson to involve your children in preparing the molds.

54 Geometry—Solid Shapes

You Will Need
- Chapter 11, Geometry
- Bring out the hardened molds from yesterday's class and cut the paper away to reveal the solid wonders!
- chart paper and markers

Your Lesson

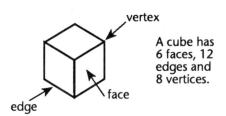

A cube has 6 faces, 12 edges and 8 vertices.

A cylinder and a cone both have circles for a base. A cone has one vertex, but a cylinder has none.

Examine each shape, looking for such things as:

- symmetry
- number of faces, edges or vertices (corners)
- right angles, parallel lines and perpendicular lines
- any similarities or differences that exist among the shapes.
- differences in volume. Which appears to have the greatest volume? You might fill some empty molds and do some pouring to find out (please see Contact Lessons 29 and 30).
- relationships with two-dimensional shapes

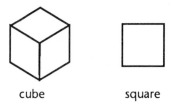

A cube has a square base.
A square has parallel sides, while a cube has parallel faces.

Record observations about these shapes on chart paper and save them for future reference.

Additional Ideas

55 Measurement—Non-Standard vs. Standard Units of Weight

You Will Need
- Chapter 7, Measurement
- several copies of a specific book, one for each group of children to weigh
  for every four children
- a ziplock bag filled with 35 pennies
- several empty ziplock bags,
- a container of filler material such as dirt, sand, or pebbles
- balance scales

Your Lesson

The purpose of this lesson is to provide experience with non-standard units of weight.

Give each group of 4 children the materials listed above. Have them fill each of the ziplock bags with an amount of dirt they feel weighs as much as their set of pennies. Ask them to suggest a name for the weight of their bags (our children suggested "Trid" for a label, which is "dirt" spelled backwards). Then have them use the balance scales to find out how many "bags" a book, such as a reading book, weighs. (Note: Each group weighs the same book.)

When all have accomplished these tasks, have the groups report the "bag" weight of the book. In all likelihood, the group weights will not agree and this should be discussed. Two explanations for the disagreement should be mentioned: Each bag probably contains a different amount of dirt. If the book doesn't weigh an exact number of bags, then a portion of a bag must be estimated.

Consider the first of these reasons. This can be tested by having each group weigh one of their bags against the bag of pennies. Whenever necessary, give them time to adjust the amount of dirt to create a balance with the pennies.

Each group now has a "standard" bag of dirt, which we will call a Trid during the rest of this lesson and Lesson 56. Using these bags, investigate questions such as:

How many Trids does above book weigh?

What is the Trid weight of other items in the room?

Suppose an object does not weigh an exact number of Trids? How might its Trid weight be reported?

55 Measurement—Non-Standard vs. ...(continued)

Save the bags of Trids for Lesson 56. Ask the children to bring some items from home that can be weighed with their Trids.

Journal Writing Ask your children to describe how their group estimated the amount of dirt needed to balance their bag of pennies.

Additional Ideas

56 Measurement—Weight, Estimation and Graphing

You Will Need
- balance scales—one for each group of four children, if possible
- Trids created from yesterday's lesson
- extra ziplocks and dirt
- items brought in from home for weighing
- 3-column graph for each child (Blackline 41)

Your Lesson

How many Trids does each item brought from home weigh?

Give each group of four a scale and a Trid of dirt. Ask them to investigate ways to estimate and determine the Trid weight of the items brought from home. As part of this, you might ask each child to indicate on a 3-column graph if their estimates were "too high", "too low", or "just right".

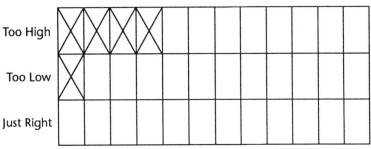

Allow time for the groups to report about their work. Did any group employ special procedures for estimating and weighing? Did any find it necessary to create additional bags that weigh a Trid? Were fractional parts of a Trid needed? How?

As time permits, have the groups switch stations and items with other groups.

Additional Ideas

57 Money—Estimation and Place Value

You Will Need
- a small, clear jar filled with coins
- chart paper

Your Lesson

At the start of the day, put your jar of coins on display, together with an invitation for the children to guess the total amount of money it contains.

When math time comes, ask volunteers to describe the strategies they used for making their estimates.

What other strategies could be used for making an estimate? Here are two possibilities:

Take one handful of coins and determine its value. Multiply this total by the approximate number of handfuls that can be drawn from the container. Does it make a difference whose hand is used?

57 Money—Estimation and Place Value (continued)

Fill a small cup with coins from the container and determine the value of those coins. Multiply this total by the approximate number of cups that can be scooped from the bowl. Note: One way to estimate the number of cups that can be scooped is to compare the weight of the coins in the container with that of the coins in the cup.

Ask the children to discuss or write about their feelings concerning these strategies. Do they notice any problems in applying them? Which strategy seems most appropriate to use? Do any estimates appear to be unacceptably large or small?

Now for the big challenge! *How much money is in the container?* Pass out the money to groups and ask them to calculate the value. Some discussion questions:

How close were the estimates? Looking back, did any one estimation strategy work better than others?

What is the difference between the closest estimate and the actual amount of money? How can this difference be shown with a physical model? with a visual model? with a collection of coins? What is the minimal collection of coins needed to show the difference? What is the minimal collection of coins needed to show the total amount of money in the jar?

Teacher Tips

It is important for children to recognize the approximate nature of any estimate and that a variety of estimates may be acceptable. It is also important that they discuss various estimation strategies and comment about the one(s) they would have confidence in using.

It would be appropriate to use a calculator throughout this activity. If necessary, multiplications can be replaced by repeated addition.

Journal Writing

If you were asked to estimate the amount of money in another jar, how would you do it? How would you feel about your estimate?

Additional Ideas

58 Money—Sorting and Fractions

You Will Need
- Chapter 2, Sorting
- a collection of 10 coins for every four children

Your Lesson

How can coins be sorted or classified?

Ask each group of four to sort their coins into two sets according to some criterion. Then have them describe their sorting criterion with words and with fractions (or other numbers).

"We separated the coins that had a 'D' (for Denver) from those that didn't; 3/10 of our coins had a 'D'."

"We found that 7/10 of our coins had a date that was an odd number and 3/10 didn't."

"We made 2 piles: coins with ridges on their edges and coins with no ridges."

As time permits, ask the groups to repeat the activity using other sorting criteria. As your children work, circulate and join in the fun with each group!

Teacher Tips

This task becomes more and more difficult as time goes on. You will find that children will begin to piggyback off each other's ideas. Before long, observations will be made that will require quite a bit of critical examination. Bravo!

Journal Writing

Complete this sentence: Today I was pleased that I _____.

Additional Ideas

59, 60 Money—Graphing

You Will Need
- Chapter 10, Data Analysis and Graphing
- a collection of 10 coins for each team of 4 children
- graphing materials

Your Lesson

In this two-day lesson, your children will gather information about coins, prepare graphs, record information on the graphs and analyze the data.

Day 1

Ask each team of four children to sort 10 coins according to some criteria, just as they did in Lesson 58.

Tell the groups that, in tomorrow's lesson, they will survey the class to find out how many children have coins that satisfy the criteria they have selected. Today each group's task is to prepare a graph that can be used to record the results of this survey. To ensure variety, assign a different graphing style to each group. An example using egg cartons is pictured here (please refer to Chapter 10 for other types of graphs).

Do you have a coin from Denver?

Place a pom-pom to show your answer.

My coin came from Denver

My coin didn't come from Denver

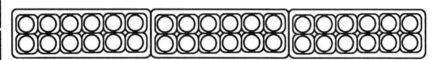

The children may need your assistance with this task. Discuss each group's plan with them and advise them where they can locate the necessary materials. Allow time for each group to set up their graph.

All is now ready for the gathering of data tomorrow!

Day 2

It is now time to conduct some surveys! Place the graphs that were prepared yesterday at various locations in the room. Give each child a coin with the instructions to visit each location and supply the information that is requested about the coin. (Note that a child keeps the same coin throughout this activity.) When all the data has been gathered, the children can return to their groups and summarize and share the data shown on their graph.

Additional Ideas

61, 62 Money—Congruence and Paper Folding

You Will Need
- Chapter 11, Geometry
- A one-dollar bill for observations *for each child*
- two rectangular sheets of paper
- several copies of Blackline 81 (Observation Record Sheet)

Your Lesson

"It is a rectangle." "Opposite sides have the same length. "There are 4 right angles." "There are 2 lines of symmetry."

Day 1

Show a dollar bill to your children and discuss its shape and symmetry.

Now conduct a paper folding activity that will enable the children to find hidden shapes within a rectangle. Have them first fold along one of the lines of symmetry as shown here.

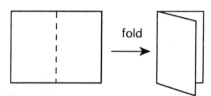

Ask the children to imagine what their paper will look like when it is unfolded. Before unfolding, have them describe what they think they will see. For example, can they predict how many rectangles and right angles will be seen? Can they predict the size of each rectangle?

Unfold the papers and check the predictions. What do the children observe?

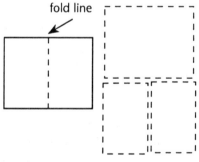

"I see 3 rectangles."
"The 2 smaller rectangles are the same size."
"Two sides of the smaller rectangles are half as long as a side of the paper."

Allow time for children to record their observations on Blackline 81.

Ask the children to fold their papers once more along the same line of symmetry. Then have them make a second fold as shown here:

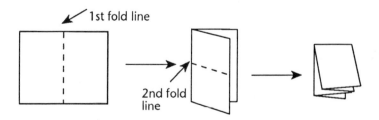

82
CONTACT LESSONS

61, 62 Money—Congruence and...(continued)

Now have each child imagine what their paper will look like when unfolded and discuss their thoughts with a partner. How many rectangles will there be? Will any of them be congruent? How many right angles will be formed?

When the children are ready, have each team unfold their paper and check their predictions. Prepare a written description of what they actually see. Reports can be shared with the class.

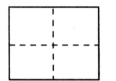

"I see 9 rectangles."
"There are 16 right angles."
"The 4 smaller rectangles are the same size."
"The smallest rectangle is $1/4$ as large as the entire page."

This activity can be extended by having the children make more folds. For example, what happens if a third fold is made at right angles to the second one?

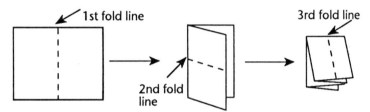

Day 2

Yesterday, the children examined the shapes created by folding a rectangular piece of paper twice. The first fold was made along a line of symmetry and the second was at right angles to the first (see diagram above). In today's activity, the first fold is made in the same way, but it is up to the children (in pairs) to make a second fold of their choice. For example, they might fold their papers like this:

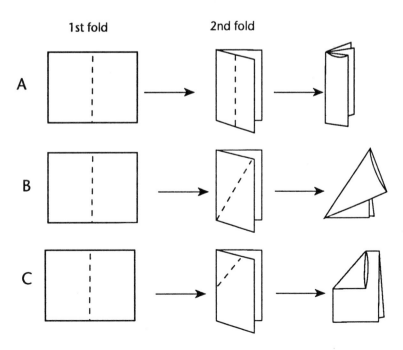

61, 62 Money—Congruence and...(continued)

As before, ask each team to predict what its papers will look like when unfolded. What shapes can be seen? Will the shapes be alike (or different) in any way? Will any be congruent? What kind of angles will be created? Will any of them be right angles? Will any be larger (or smaller) than a right angle?

When ready, each team can unfold its paper, check its predictions and prepare a written description of what they actually see. Reports can be shared with the class.

Teacher Tips You may wish to make an art lesson out of this activity. Your children will enjoy coloring the shapes they have folded.

Journal Writing Describe how you felt about the paper-folding activities you did today.

Additional Ideas

63 Money—Geometry

You Will Need
- Chapter 11, Geometry
- individual copies of Blackline 75 (Circle)
- scissors

Your Lesson

All U.S. coins are circular. Today's activity examines some of the shapes that can be folded from a circle. The activity is adapted, with permission, from *Visual Mathematics, Course I*, by Foreman and Bennett (©Math Learning Center, 1991).

Instruct each child to carefully cut around the circumference of their circle and to then fold point A over until it just covers the center dot. Do the same with points B and C.

What shape was formed from these fold? Unbelievable—eh! The children will even fold and unfold repeatedly to experience the magic of a circle transforming into a triangle. What observations can be made about this triangle? If care has been taken, the triangle will be equilateral, with all sides the same length and all angles the same size. Ask the children to suggest ways of informally verifying these results.

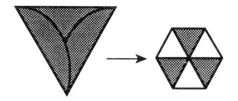

Now ask the children to fold each vertex of their equilateral triangle into the center dot.

Now what shape has been formed? How many sides and angles does it have? Are the sides congruent? How can this be tested? How many equilateral triangles can be seen in this shape? What other things can be observed about this shape?

Teacher Tips

In Part 2, the shape formed should be a regular hexagon, where all sides and angles are congruent.

Provide extra copies of Blackline 61 to share with the folks at home.

Additional Ideas

64, 65 Money—Story Problems

You Will Need
- classified ads from the newspaper

Your Lesson

Day 1

The classified ads are full of items of interest to young children. For example, we know a family whose youngest member learned to read the classified ads at school. There was a "Schnocker" for sale in the local paper for a mere $30. What is a Schnocker? Why it is a puppy whose mother was a Schnauzer and father was a Cocker. By the time that child's parents arrived home from work, all the children in the family had ganged up and, you are right, pooled their money. By bedtime, one perfectly adorable Schnocker had a new home, much to the joy of the children!

What items of interest can your children find in the local classified ads? Challenge them to look carefully for the bargain of the century. Have them write story problems about the ads. Here are a few we have received.

There were 3 kittens for sale. Each kitten cost $45. How much money would be needed to purchase all 3 kittens?

Mrs. Jones will cover a chair for $55 and a sofa for $125. How much money would be needed to pay Mrs. Jones to cover 1 sofa and 3 chairs?

A Siamese kitten costs $225 and a Persian kitten costs $135. How much more money is needed to purchase a Siamese kitten?

Ask the children to attach their ads to their stories and save them for the next day's class.

Day 2

Working in small groups, challenge your children to read and solve the problems that were written yesterday by their friends. Encourage them to estimate answers and to use sketches (or number pieces) to help them construct solutions. After an appropriate time, have volunteers from each group share their strategies for solving some of the problems.

Teacher Tips

These activities can help children strengthen their number sense. Encourage the groups to discuss how they will calculate answers. Can they use mental arithmetic? Would calculators be appropriate?

Additional Ideas

66 Mystery Box—Flags; Geometry—Symmetry

You Will Need
- Chapter 11, Geometry
- Mystery Box containing a flag of any type
- translucent money
  for each pair of children
- an 8½ × 11 piece of paper
- markers such as crayons or colored pencils

Your Lesson

The theme of flags will begin in a few days (Lesson 74) so begin to prepare for it with a mystery box activity. Begin the questioning process sometime during this day and see how long it takes the children to guess that there is a flag inside!

One way to conduct the activity is to allow the children a dollar's worth of questions each day, with each question costing 5¢. Begin by displaying a translucent dollar bill on the overhead. Then, after each question, a volunteer removes 5¢ from the overhead, trading coins as needed.

In earlier Contact lessons, symmetrical designs were created from pattern blocks and from apple shapes. Refresh your children's memories of these experiences and review the meaning of a line of symmetry by referring to symmetrical objects. Discuss the fact that such an object can be divided into two parts that are mirror images (or reflections) of each other.

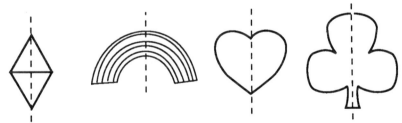

Have children form pairs and ask each team to divide its paper into two parts with a fold. Tell them the crease is to be a line of symmetry. One member of each team is to create a design on one side of the line of symmetry. The other is to draw the mirror image of the design. Then have partners reverse roles.

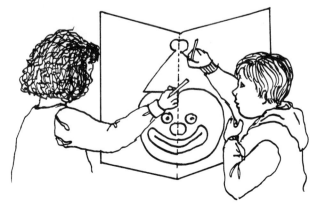

66 Mystery Box—Flags; Geometry...(continued)

As time permits, teams can repeat the activity with the children reversing their roles.

Some children may choose to create abstract designs while others work together to create a recognizable one. In any case, allow time for them to describe their pictures and to comment about the location of each part of the design. The line of symmetry should separate the design into two parts that are mirror images of each other. If the paper is folded over on the line of symmetry, the two parts of the design coincide.

Journal Writing Our children enjoyed writing descriptions of their creations in their Math Journals. They incorporated lots of mathematical language. Give your class this writing assignment.

Homework (Optional) Have your children return copycat drawings that have been created with someone at home. Display them for all to enjoy. They can also bring in symmetrical designs or shapes they find in their homes.

Additional Ideas

67, 68 Geometry—Congruence

You Will Need
- Chapter 11, Geometry
- a transparent geoboard or a transparency of Blackline 68 *for each child or small group*
- geoboards,
- rubber bands
- Blackline 68 (Geoboard Recording Paper)

Your Lesson

In this two-day lesson, children use geoboards to examine congruent segments and regions.

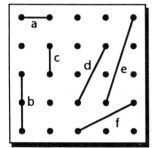

Day 1

Display the following segments on a transparent geoboard at the overhead. Ask the children to form these segments on their boards.

If your children have difficulty with this task, help them with statements such as, "Form segment **d** by going down 2 spaces and over 1."

Ask the children to compare the lengths of these segments. Do any have the same length? Is **b** longer than **d**? How about **d** and **f**? etc. Let children create their own methods of comparison. Some may use their index fingers, while others may use strips of paper or rulers. Invite volunteers to describe how they arrived at their conclusions.

Don't miss the opportunity to develop appropriate language. For example, in the above exercise, segments **a** and **c** are the same length. Therefore, they are congruent.

Continue the lesson by asking the children to compare the lengths of other geoboard segments. You might also allow time for the children to work in pairs. One member of the pair makes a segment on their board and challenges the other to make a segment that is congruent to it.

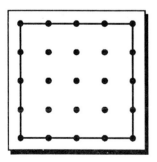

Day 2

This is one your children are sure to enjoy! Show the class the square to the left and ask the children to make it on their boards.

Place a second rubber band as shown to the right and have the children do the same.

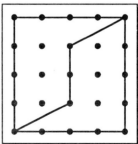

Does it seem to the class the square has been split into two regions that have the same size and shape? Have the children verify this by recording the shapes on geoboard paper, cutting them out and placing one on top of the other. These regions are examples of congruent regions.

67, 68 Geometry—Congruence (continued)

Ask the children to start with the original square once more and to find other ways to split it into two congruent regions. There are many ways to do this, some of which are pictured below. Have the children record their solutions on geoboard paper. If unsure of a solution, they can cut it apart and see if one part will fit exactly onto the other.

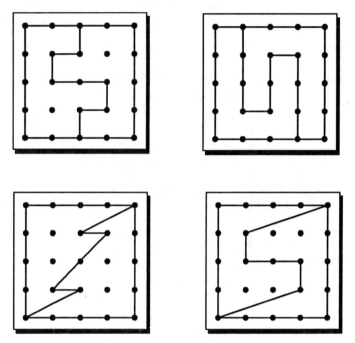

Teacher Tips Children enjoy playing "copycat" with their geoboards, where one child creates a shape and asks a friend to copy it on another board. This always seems to lead to personal challenges in our classroom.

Journal Writing What would you say or draw if you wanted to explain the word "congruent" to a friend?

Additional Ideas

69 Measurement—Volume

You Will Need
- Chapter 7, Measurement
- some open-topped boxes (about the size of a small cereal or shoe box)
for each child
- ¾" wooden cubes
- a copy of Blackline 81 (Observation Record Sheet)

Your Lesson

Show one of the boxes to the class and discuss strategies for answering the following question: Approximately how many cubes will be needed to fill this box? Have the children record their thoughts on their Observation Record Sheets.

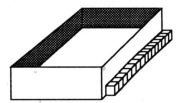

"About 15 cubes can be layed against this edge. I think it will take about 20 of these rows to fill the box. So my guess is 15 x 20 or 300 cubes are needed."

Compare the estimates that are made with the actual volume of the box by asking the children to actually fill the box with as many cubes as it will hold.

Discuss such questions as:

Assuming that the box was not exactly filled with whole cubes, how can the left over part be filled?

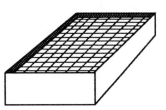

"It looks like you could fill the box with a row of cubes that have been split in half."

"We could replace our cubes with ones that are smaller."

If smaller cubes are used to fill the box, would we need more or less of them?

Do wider boxes always hold more? Why or why not?

If we wanted to tell a friend how many cubes were needed to fill a box, would consistency of cube size be a factor to consider? (This is an opportunity to discuss the use of standard units of volume.)

Allow time for the children to complete their record sheets.

As time permits, repeat this activity using other boxes.

Additional Ideas

70 Spatial Relationships—Cube Covers

You Will Need
- Chapter 11, Geometry
- ¾" wooden cubes
- 2-cm grid paper (Blackline 73)
- scissors and tape

Your Lesson

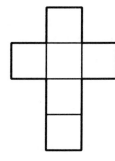

What 2-dimensional patterns can be folded to exactly cover a cube?

Begin an investigation of this question by asking your children to cut the pattern of squares (shown at the left) from their grid paper.

Demonstrate how this pattern can be folded to exactly cover each face of a cube. Ask them to cover their cubes in the same way.

Challenge the children to find other patterns that are "cube covers". There are several that will work, such as these:

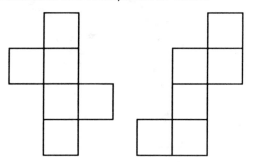

Children will likely conduct this search in various ways: guess-and-check, cutting and taping, rolling a cube onto the squares, or attempting to mentally visualize a cube unfolding into a pattern.

Display the patterns that are found and discuss the children's observations. Do they notice any similarities or differences among the patterns? In patterns such as those below, do the children consider them to be the same or different?

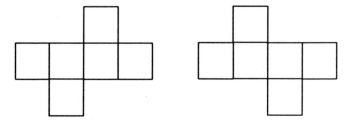

"Are these really the same pattern?"

Additional Ideas

92
CONTACT LESSONS

71 Measurement—Surface Area

You Will Need
- Chapter 7, Measurement
- containers of wooden cubes

Your Lesson

In Contact Lesson 70, your children covered a cube with various patterns of 6 squares. Each of the squares corresponds to a face of the cube; therefore the surface area of the cube can be reported as 6 square units. (This assumes that each square represents 1 square unit of area, an assumption made throughout this lesson.) When discussing this with children, we also like to use the image of a cube "floating in space". One can count 6 square faces on such a cube, and so its surface area is 6 square units.

Today's lesson focuses on thinking visually about the surface area of small collections of cubes. Here is one way to proceed.

Give each child a collection of 6 to 8 cubes and discuss the following questions: How many square faces would be exposed if 1 cube were floating in space? (Answer is 6.) If 2 cubes were floating in space, how many square faces would be exposed? (More than one way of thinking may be shared.)

How many square faces would be exposed if 5 cubes were floating? 8 cubes?

Place 1 cube on a surface. How many square faces are exposed? (5) How many are exposed if a second cube is placed on the surface?

Divide the children into pairs. Give each pair 20 wooden cubes and pose the following problem:

Suppose the cubes were placed end-to-end in a line to form a train, how many square faces would be exposed altogether?

71 Measurement—Surface Area (continued)

Have children work together to determine the total. Allow ample time for sharing. You may have such suggestions as:

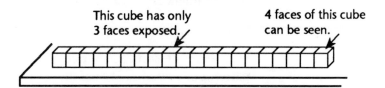

"Four square faces are exposed on each end cube. That's 8. Each of the other 18 cubes has 3 faces exposed. That makes 18 × 3 = 54 more. Altogether there are 54 + 8 = 62 exposed square faces."

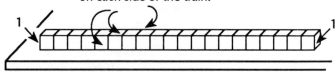

"Altogether, there are 20 + 20 + 20 + 1 + 1 = 62 exposed squares."

Have the children work in pairs to create a 4 × 5 rectangular array of cubes. How many square faces are exposed altogether? Let your children share with the entire class how they discovered the total. Expect such answers as:

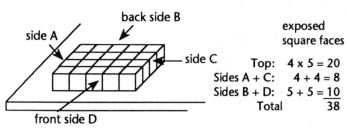

	exposed square faces
Top:	4 × 5 = 20
Sides A + C:	4 + 4 = 8
Sides B + D:	5 + 5 = 10
Total	38

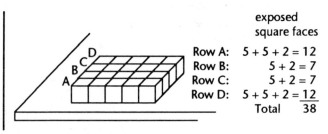

	exposed square faces
Row A:	5 + 5 + 2 = 12
Row B:	5 + 2 = 7
Row C:	5 + 2 = 7
Row D:	5 + 5 + 2 = 12
Total	38

Others may count and touch each surface individually.

If the 4 × 5 array were floating in space, how many square faces would be exposed?

Emphasis

The number of square faces that are exposed in an arrangement of cubes depends on the arrangement: If the cubes are floating individually in space, each cube has 6 exposed square faces. The number of exposed square faces is fewer when a cube is placed on a table or when cubes are placed side-by-side (as in a train or array).

Additional Ideas

72 Measurement—Surface Area

You Will Need
- Chapter 7, Measurement
- 100 wooden cubes for every four children

Your Lesson

How many square faces are exposed when 100 cubes are arranged in different rectangular arrays?

Give each group 100 wooden cubes, and review some of the ideas discussed in Contact Lesson 71. If all cubes were floating individually in space, how many faces would be exposed? How many square faces would be exposed if each cube sat alone on a surface?

Ask each group to form a 1 × 100 array on the floor. How many square faces are exposed?

Have each group brainstorm together to determine an answer to this question. Some generalizations and predictions will result from the varied ways groups think about this problem. Some of the methods that might be used:

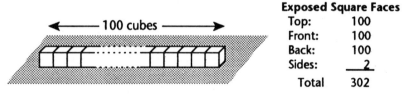

Exposed Square Faces
Top: 100
Front: 100
Back: 100
Sides: 2
Total 302

Exposed Square Faces
2 end cubes: 4 + 4 = 8
middle 98 cubes: 98 × 3 = 294
Total 302

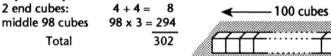

Allow time for each group to share their strategies with the class.

Now ask each group to form a 2 × 50 rectangular array on the floor. How could the exposed square faces be counted? Illustrated below are two ways the children might arrive at their total.

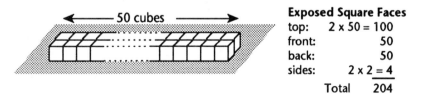

Exposed Square Faces
top: 2 × 50 = 100
front: 50
back: 50
sides: 2 × 2 = 4
Total 204

72 Measurement—Surface Area (continued)

There are 48 sets of 2 cubes in the middle. Each set has 4 square faces showing. That's 48 x 4 = 192 square faces.

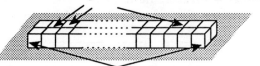

These end cubes each have 3 square faces exposed. That makes 4 x 3 = 12 squares.

Altogether, there are 12 + 192 = 204 squares showing.

Continue by asking the groups to form other arrays with their cubes. Ask them to discuss strategies they could use to determine the number of exposed faces in each case.

Close your lesson by connecting strategies learned today with architecture. How does a bricklayer know how many bricks will be needed? Must one count every brick?

What connections can be made to the human body? Do "exposed surfaces" play a part when we try to keep warm? Why do we curl up in bed when we're cold? When we have fewer exposed surfaces to the cold, are we warmer?

Emphasis   Children may total the number of exposed faces in an arrangement of cubes in several ways. Some may attempt to count each face one at a time, while others may use visual patterns in the arrangement to help them determine an answer.

Journal Writing   Write a brief paragraph that describes how you would total the number of exposed faces in a train of 14 cubes.

Additional Ideas

73 Fractions—Fraction Bar Wall Chart

You Will Need
- Chapter 8, Fractions
- A set of Group Fraction Bars for every four children
- a chart to be displayed
- a set of Group Fraction Bars to be used for display

Your Lesson

Earlier in the year, your children used Group Fraction Bars to show information about themselves (see Contact Lessons 33–35). In this lesson, a display will be prepared showing the fractions represented by these bars.

Give each group of four a complete set of Group Fraction Bars and ask the groups to sort the bars by color.

Re-examine the bars of each color, discussing the fractions that are represented. As you do so, post a sample of each bar on the class chart as shown below. Invite your children to make observations or write questions about these fractions. Record these on conversation bubbles posted around the display.

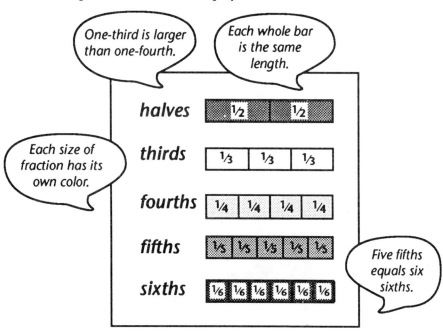

Note: Each type of fractions bar is cut from a single rectangle (whole bar). The parts are glued close together so students can easily see that all the whole bars are the same size.

Display the chart for the rest of the year and use it as a reference.

Additonal Ideas

74, 75 Fraction Bar Discoveries

You Will Need
- Chapter 8, Fractions
- a set of Fraction Bars for each child (purchased or made from Blacklines 90–94)
- overhead Fraction Bars
- the Fraction Bar Wall Chart from Lesson 73

Your Lesson

Day 1

In this lesson, your children will examine a complete set of Fraction Bars for the first time. These sets include bars that show halves, thirds, fourths, fifths, sixths, tenths and twelfths. They may be purchased from The Math Learning Center or duplicated on construction paper. (See making instructions in the Materials Guide.)

Begin by displaying a transparency of the green bars at the overhead. Ask the children to describe each bar, name the fraction it shows and then find the bar that matches it in their set. Record the appropriate fraction beside each bar and discuss the meaning of the numerator (the number of shaded parts) and the denominator (the number of equal parts in the bar).

Notice that some of these bars are also pictured on the wall chart that was prepared during Contact Lesson 73.

Go through the same procedure for the bars of each color. As the bars are examined, ask the children to share any observations they have.

Each color has a bar with no shading.

Each color has a bar that is completely shaded.

All the bars are formed from rectangles that are the same size.

The more sections a bar is divided into, the smaller the sections are.

All the green bars are divided into 2 equal parts. All the yellow ones show thirds...

There is always one more bar than the total number of parts in each bar. Three bars for halves, 4 bars for thirds...

74, 75 Fraction Bar Discoveries (continued)

Day 2

When using Fraction Bars, the denominator of a fraction is represented by the number of parts in a bar and the numerator by the number of shaded parts. The unit is the length of an entire bar. This model for fractions is helpful for thinking about equal and unequal fractions in a visual manner (please see the dialogue on page 73 in the Teaching Reference Manual).

Spend this day using the bars to promote a discussion of how fractions are related to one another. Exercises such as the following would do this:

Hold up a red 2/6 bar and ask the children to find other bars that have the same amount of shading. Describe the fractions that are found.

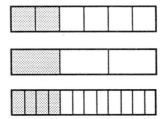

Since the bars have the same amount of shading, we say their fractions are equal. Thus, $2/6 = 1/3 = 4/12$.

Search for other sets of equal fractions in the same way. You might even have the children sort their Fraction Bars into piles according to shaded amounts. Discuss the methods they use for deciding if two bars have the same amount of shading.

Hold up two bars that have different amounts of shading. Ask the children to describe the corresponding fractions and to decide which fraction is greater. Have them share their thinking in a show-and-tell manner.

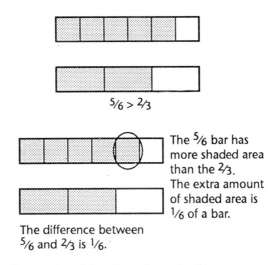

Compare other pairs of fractions in a similar way.

76 Flags Sorting—Venn Diagrams

You Will Need
- Chapter 2, Sorting
  for every group of four children
- a set of flags cut from Blacklines 71 and 72
- two loops of string

Your Lesson

You may wish to begin this lesson by discussing the history and features of the American flag or of your state flag.

Divide the class into groups of four and distribute the above materials. Ask the groups to place their flags inside the loops in a way that separates the flags into two different sets. Have them give each set a title that describes how it was formed.

Have the groups share some of their sorting procedures. If some groups created overlapping sets, these may need to be discussed.

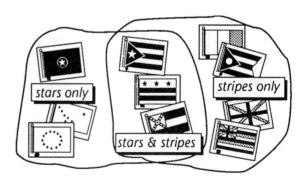

Teacher Tips

This lesson can be varied by specifying the sorting criteria. For example, ask the children to devise a way to place the flags with stars in one loop and those with stripes in the other. In this case, as shown above, the loops will have to overlap in order to represent both sets.

76 Flags Sorting—Venn Diagrams (continued)

Another variation is to graph information about the flags. One possibility might be to have each group randomly select 20 flags and make a graph highlighting some of their features.

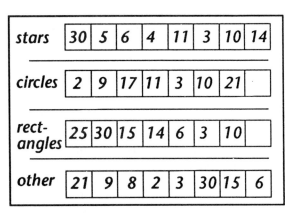

The numbers listed here refer to the flags on Blacklines 71 and 72.

How many flags appeared in all 4 rows? In 3 of the rows?

Does any row have twice as many flags as some other row?

What fractional part of the 20 flags have rectangles?

Is it likely that all groups contain the same flag?

Encourage children to use fractions to describe their sorted collections.

Journal Writing Write a paragraph that describes one way your group sorted its flags.

Additional Ideas

77 Flags—Patterns

You Will Need
- Chapter 3, Patterns
- a copy of Blackline 69 (United States flag) for each child
- scissors
- transparency of Blackline 77 (Folded flag)

Your Lesson

What is the correct method of folding Old Glory? What shape is formed by the folded flag?

Your class might research these questions in an encyclopedia or a scout handbook and practice the required folding technique with a school flag or a beach towel.

Have the children cut out the flag on Blackline 69 and fold it using their new folding skill, being very careful as they proceed. Ask them to predict, in writing, what pattern of shapes will be observed when the flag is unfolded. Discuss their thoughts.

The students can now check their predictions by unfolding their papers and observing the pattern of triangles on the back. Display this pattern on the overhead and invite the children to identify some of the shapes.

Discuss this pattern further. How was it determined by the method of folding? How are the triangles in the pattern related? Are they the same size? Do they have the same shape? Are they different in any way?

Conclude by having the students write summary statements about the pattern.

Emphasis

The triangles that have been formed by successive folds should be congruent to one another. That is, they have the same size and shape. However, those that are reflections of one another will have different orientations.

The pattern of triangles is an example of a tessellation. Please see pages 91–92 of the Teaching Reference Manual for more information about tessellations.

77 Flags—Patterns (continued)

Teacher Tips For an art lesson, or optional homework, have children color their folded flag paper of triangles. Display their pictures for all to see. Illustrated below are a few examples.

Homework The folding method used in this lesson generated a pattern of triangles when the flag was unfolded. Challenge the children to devise folding methods that will produce patterns of rectangles or other shapes. Ask them to share samples of their work during the next lesson.

Additional Ideas

78 Flags—Estimation, Measurement

You Will Need
- Chapter 7, Measurement
- a rectangular flag (18 × 24 or larger)
- square tile (units from the base ten pieces work nicely)

Your Lesson

Show the class a flag and some square tile. How can the number of tile needed to cover the flag be estimated? That is, what is the approximate area of the flag?

As in previous estimation lessons, give the children an opportunity to make suggestions for answering the above question. Some possibilities:

Make a guess after imagining the flag being covered by the tile.

Determine how many tile will fit across the flag in a single row. Multiply this number by the approximate number of rows that can be made.

Find out how many tile will cover a smaller rectangle, such as a book. Multiply this total by the approximate number of similar-sized books that will cover the flag.

Determine an estimate for each of the suggested strategies.

How do the estimates compare?

Which strategy seems most appropriate?

Check the estimates by actually "tiling" (covering) the flag with the squares. One way to total the number of squares is to group them into mats, strips and units. Suppose the squares do not exactly cover the flag? How can a total be reported? What suggestions do your children have?

Teacher Tips

This lesson can also be done with other materials, such as hex-a-links or pattern blocks.

You might want to extend (or replace) this lesson by having your children cover a blackline of the U.S. flag with base ten pieces. If so, here are some questions that can be discussed:

How many units are needed?

What is the minimal collection of pieces needed? What is the maximum collection? Is the area the same in each case?

What if the flag's area is doubled? Challenge your children to visualize it in their mind's eye. What would the new area be?

Additional Ideas

79 Flags—Measurement

You Will Need
- Chapter 7, Measurement
- pattern blocks for every four children
- pattern blocks for the overhead

Your Lesson

Divide the class into groups of four and ask each group to:

Create a star with its pattern blocks.

Measure the perimeter of its star, using the *side* of the green triangle as the unit.

Make other stars with pattern blocks and find the perimeter.

End your lesson with observations made by the children. Was the perimeter of every star the same number of units? Why or why not? How did the perimeter change?

Teacher Tips

The children might determine the perimeters of other shapes made from the pattern blocks. Challenge them to make a rectangle with a perimeter of 16 units. How many different rectangles with a perimeter of 16 units can they make? Can they create any with a perimeter of 20 units? How about some having a perimeter of 15 units?

Additional Ideas

80, 81 Flags—Story Problems, Money

You Will Need
- a supply catalog that advertises your state flag
- base ten pieces, grid paper and calculators, as needed

Your Lesson

Day 1

How much does your state flag cost? Look in a school supply catalog and ask the children to describe different combinations of money that will pay for it.

Have children work in pairs to create story problems about flags. Encourage them to include the price of flags, if possible. Here are some story problems our children have written:

• A large state flag costs $19.99. A smaller one is $6.99. If a class had $25 to spend, how many large flags could it buy? How many small ones? How many of each for $50? $100?

• If each flag cost $9.98, how much money would it cost to buy one for our state and every bordering state?

• Todd, Jeanie, Shanda and Susanna each have U.S. flags at home to fly for holidays. Each flag has 50 stars. How many stars in all are on the 4 flags?

• Edward and Peter wanted to decorate the yard for their family's 4th of July party. They decided to hang USA flags from the trees. They needed 23 flags to really look good. Each flag cost $1.99. How much money did they need to buy the flags?

• If Betsy Ross sold her flags today for $50 each, how much money would she get for 2 flags? 10 flags? 30 flags? If her profit for each flag were $14, how many would she have to sell to make $100 profit?

Ask the groups to save their problems for the next day's class. Also, as time permits, have them prepare a separate answer sheet.

Day 2

Divide the class into the same groups as yesterday. Have them exchange their problems and begin solving the ones they receive. After an appropriate time, have volunteers from each group share their thinking when solving some of its problems.

Teacher Tips

This lesson can help children strengthen their number sense. Encourage them to estimate answers and to decide when it might be helpful to use mental arithmetic, diagrams or calculators.

Journal Writing

Describe how you felt about the work you did in your group. Was any problem especially easy for you? Did any cause you difficulty?

82 Flags—Probability

You Will Need
- Chapter 9, Probability; Chapter 2, Sorting
- a set of flags, 1–15 (cut apart Blackline 71), and a paper sack
- transparency of Blackline 74 (Flag Probability)

for each group of four children
- two loops of yarn and a paper sack
- a set of flags, 1–15, (cut apart from Blackline 71)
- a second set of flags, 1–15, cut apart, and placed in the sack
- a copy of Blackline 74

Your Lesson

Today's lesson explores both probability and sorting. Distribute two loops of yarn and a set of flags, 1–15, to each group of four. Ask the groups to sort their flags, placing those with circles within one loop of yarn and those with stars within the other. The loops will need to overlap, since some flags have both circles and stars. Also, flags that have neither circles nor stars will lie outside the loops. A complete sorting is indicated here.

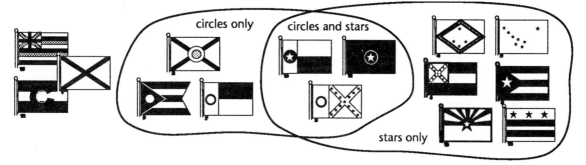

In front of the class, place your set of flags, 1–15, in a paper sack and thoroughly mix them. Now conduct the following probability activity.

Display a transparency of Blackline 74 and randomly draw a flag from the sack. On the transparency, indicate with a tally the category into which the flag falls and put the flag back into the sack. Do this a few more times. Each time, examine the drawn flag, make a tally in the appropriate category on the transparency, and then put the drawn flag back into the sack. A possible sequence is illustrated here.

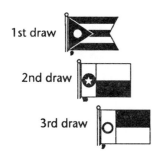

1st draw
2nd draw
3rd draw

Flag Probability Prediction: Prediction:			
Circles Only	Stars Only	Circles and Stars	No Circles and No Stars
//		/	

107
CONTACT LESSONS

82 Flags—Probability (continued)

Ask the groups to imagine that 20 draws will be made in the manner described above. On Blackline 74, have them record predictions to the following questions: How many times do they think a flag that has both circles and stars will be drawn? Why? Which will be drawn more times: a flag with a star or a flag with a circle? Why?

Discuss their predictions. Then distribute sacks containing flags, 1–15, to the groups and ask them to tally the results of 20 draws on their blacklines. Have them summarize the results.

Allow time to conduct a show-and-tell discussion of each group's work.

Teacher Tips

Theoretically, a flag with both circles and stars should be drawn $\frac{1}{5}$ of the time. This is because 3 of the 15 flags fall into this category, as seen in the Venn Diagram at the start of the lesson. Therefore, one might expect this kind of flag to be drawn about 4 times out of 20.

Similarly, there are 9 flags with a star and 6 with a circle, so those with a star should be drawn more often.

Actual results may vary, however, because 20 draws may not be a large enough sample or because groups may not make their draws randomly.

If time permits, it would be helpful to pool the group results and compare these to what should occur theoretically.

Journal Writing

Complete the following sentence: Today I was surprised by _____.

83 Flags—Extended Number Patterns

You Will Need
- Chapter 3, Patterns
- a transparency of Blackline 70 (Four Number Flag)
- a copy of Blackline 70 for each child
- materials for making and displaying "four" flags

Your Lesson

Have your children spend some time reading about signal flags and their different uses. They will find, for example, that some flags convey warning signals and others represent numbers. The designs on number flags are especially interesting to observe and can be used to highlight various number patterns.

Discuss the features of the "four" flag with the children.

Notice that it has 4 dark sections that are separated by 2 white perpendicular lines. Have the students make copies of this flag and "hoist" them in a manner that shows multiples of 4 sections. One way for doing this is shown here.

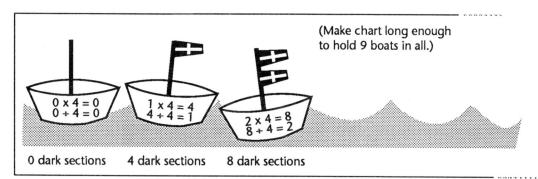

Additional Ideas

84–88 Exploring the 4 Counting Pattern

Your Lesson

The activities of Contact Lesson 83 provided a context for thinking about the 4 counting progression 4, 8, 12, 16, 20, 24, etc. The main objective of the next few days is to explore this numerical pattern further. To do this, follow the general procedures used to explore the 2, 5 and 3 counting patterns in earlier lessons (see, for example, Contact Lessons 20–24). These procedures are also summarized in the booklet *Opening Eyes to Mathematics Through Musical Arrayngements*.

Teacher Tips

This exploration begins with modeling the 4 counting progression with rectangular arrays that have a dimension of 4. It is helpful to connect this activity with those of Lesson 83 with questions such as:

Which array could be used to picture the total number of dark sections contained in 6 of the "four" number flags?

How many "four" flags are needed in order to show a total of 32 sections? Which array pictures these sections? How many flags are needed to show to 56 sections? etc.

There are many patterns children will notice when the multiples of 4 are covered on a hundreds matrix.

1	2	3	☐	5	6	7	☐	9	10
11	☐	13	14	15	☐	17	18	19	☐
21	22	23	☐	25	26	27	☐	29	30
31	☐	33	34	35	☐	37	38	39	☐
41	42	43	☐	45	46	47	☐	49	50
51	☐	53	54	55	☐	57	58	59	☐
61	62	63	☐	65	66	67	☐	69	70
71	☐	73	74	75	☐	77	78	79	☐
81	82	83	☐	85	86	87	☐	89	90
91	☐	93	94	95	☐	97	98	99	☐

"The rows go back and forth—first 2 numbers are shaded, then 3, 2, 3, etc."

"When I count by 10s, only 20, 40, 60, 80 and 100 are covered. If I continue, I predict the next ones will be 120 and 140."

84–88 Exploring the 4...(continued)

Observations such as these are helpful for predicting if larger numbers will fit the counting pattern. Also, calculators can be used to check predictions.

Complete the exploration of the 4 counting pattern with some calculator activities. Ask questions such as:

How many 4s make 60? Make a prediction and check it with a calculator, perhaps by using repeated addition. Draw a rectangle on grid paper that shows this answer.

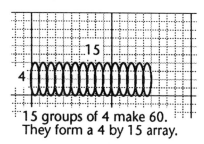

15 groups of 4 make 60.
They form a 4 by 15 array.

Do the same for other numbers. How many 4s make 96? 136? etc.

Imagine (or sketch) a rectangle that has 4 rows of 24 square units. How many square units are there altogether?

Imagine (or sketch) a rectangle that contains 104 square units and that has 4 rows. What is the other dimension? How can a calculator help here?

Be sure to allow time for the children to share their observations and thinking about these questions.

Additional Ideas

89 Measurement—Weight

You Will Need
- Chapter 7, Measurement
- an unopened box of graham crackers
- a balance scale
- egg cartons and pom-poms (or other graphing materials)

Your Lesson

This lesson begins a measurement unit that focuses on a box of graham crackers. During the next several days, your children will strengthen their understanding of weight, length, area and volume by measuring a box of crackers in different ways. The major question for today is, "How heavy is a box of graham crackers?"

Pass the box of crackers among your children so they can feel how heavy it is.

Ask the children to search throughout the room for an object they think weighs the same as the box. When they have done this, have them use a balance scale to compare the weight of their items to the box's weight.

Record the results of these comparisons graphically and invite the class to make observations about the data. An egg carton graph might look like that shown to the left.

(Graph labels: "My guess was too heavy", "My guess was too light", "My guess was JUST RIGHT!")

"14 of us chose items too heavy!"

"Only 3/21 had items that balanced with the cookies."

"An even number of people had items too light."

Discuss the children's observations. How many of the items were heavier than the actual weight of the box? How many were lighter equal to the weight?

It's delightful to see anticipation on children's faces as they come to weigh their items against the graham crackers. Hope often turns to surprise when they see how quickly one side of the scale falls! Sometimes the size of the box deceives children and makes its contents appear heavier or lighter than it actually is.

Teacher Tips

If time permits, allow your children to select a second item for comparison. Were they more successful this time?

You may wish to have each child graph the results at their desks.

Additional Ideas

90 Volume—Estimation

You Will Need
- Chapter 7, Measurement
- an unopened box of graham crackers
- a scale
- napkins

Your Lesson

How many graham crackers does your box contain?

Show your children the box of graham crackers, and ask them to estimate an answer to the above question. They may discuss whether they will estimate the number of whole crackers or the number of halves (or even fourths).

After they make several guesses, ask the children to discuss how they might refine their estimates. What information might help them? They may request to see a whole package (or a small pile) of crackers and follow strategies such as:

Multiply the number of crackers in a package (or pile) by the approximate number of packages (piles) that might fill the box.

Estimate the number of packages that will fill the box by comparing the weight of one package with that of the entire box. This number can then be multiplied by the number of crackers in a package.

Record the estimates and then determine the actual number of crackers (perhaps by multiplying the number of packages in the box by the crackers in one package). Discuss questions such as:

What was the range of guesses (that is, the difference between the highest and lowest estimate)?

How did the estimates compare to the actual amount? How many crackers "off" was the closest estimate?

If some of the estimates were rounded to the nearest 10, what totals would result?

How many crackers might we expect to find in 2 boxes? in 10 boxes? etc.

How many half crackers are in the box? How many quarter crackers are there?

Will the number of crackers in a box vary from box to box? Why or why not?

Teacher Tips

The multiplications described above can be computed by repeated addition, if necessary.

We also allow time for the children to determine the difference between their estimates and the actual count. We found this to be a good mental arithmetic activity.

91 Geometry—Spatial Visualization

You Will Need
- Chapter 11, Geometry
- several empty graham crackers boxes
- tape and scissors for every group of four children

Your Lesson

What patterns of rectangles can be formed by unfolding a graham cracker box?

Do your children remember making cube covers from squares in Lesson 70? Today's lesson is similar, only this time the children will investigate how to cut a graham cracker box so it can be unfolded into a pattern of rectangles.

Give each group of four an empty graham cracker box. Ask the groups to complete the following tasks:

Tape together the flaps on the open end of the box so as to form a rectangular prism.

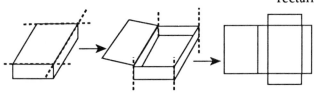

Devise a way of cutting the box along its edges so that it may be unfolded into one flat piece of cardboard. Several patterns of rectangles are likely to be created, such as those shown to the left.

Examine their patterns for a few minutes and record their observations.

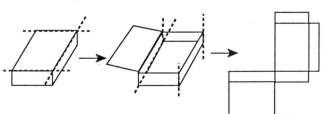

Allow time for the groups to describe their procedures for cutting their boxes and to display their patterns. Have them share their observations about the patterns and discuss these questions.

How many rectangles are part of each pattern?

How many different sizes of rectangles are seen?

How many square corners can be counted within each pattern?

Suppose the ends of the box were hexagons instead of rectangles, how many sides would the box have?

Why is a graham cracker box shaped like a rectangular prism?

Teacher Tips

Manufacturers usually consider several factors when deciding on the size and shape of containers used to package products. You may wish to discuss some of these. For example, why are some containers shaped like rectangular prisms? Why are others cylindrical (or some other shape)?

Note: Save cut boxes for Contact Lesson 93.

Additional Ideas

Insight Lessons
1 – 67

1 Getting to Know You—People Sorting

You Will Need
- Chapter 2, Sorting

Your Lesson

(1–2 days) A fun way for your students to get to know one another is to sort the class into two groups, using different attributes (for example, color of hair) as sorting criteria.

A favorite procedure of ours is to play Silent Sam with the children. Whenever Silent Sam is drawn on the chalkboard, nobody is permitted to speak.

Decide on an attribute and with exaggerated clicks of your fingers and charade-like mimes, signal a few children who have that attribute to form a group in front of the class. In the same way, direct some who do not possess it to form a second group. Involve the students in this by having them point out where they think a selected child should go. After an appropriate time, erase Silent Sam and invite the children to share their observations. Can they identify the sorting criteria and direct those at their seats to the correct group? If not, draw Silent Sam again and continue the process.

Teacher Tips

Sometimes, we vary the procedure by letting a student assume the role of the teacher. The children love to sort their friends! In fact, some will spend a great deal of time trying to devise a sorting plan that will stump their classmates.

Have the children summarize the results of each sorting.

Additional Ideas

2 Getting to Know You—Estimation

You Will Need
- individual chalkboards, chalk and erasers
- hex-a-link cubes

Your Lesson

(1 day) Leave your vanity at home for this lesson! Use each person's age, including yours, as a context for helping children think about numbers and place value.

Here are some ideas for doing this:

Have the students record estimates of your age on their chalkboards. (These can then be placed in order, from lowest to highest, in the middle of your floor area.)

What is the range of estimates? What odd numbers can be spotted? Could any be added together to produce an answer that is close to 70? To the nearest 10, which of the guesses would round to 30?

Take a deep breath, reveal your age and have a volunteer put out 1 hex-a-link cube for each year. How should the cubes be linked so as to make it easy to count them? The children will suggest different ways, perhaps by making as many stacks of 2 or of 5, as possible. Model each suggestion, being sure to include grouping the cubes by 10s and 1s.

Have each child represent their age with hex-a-links and place the towers in the floor area. What is the combined age of all the children? Answer this by arranging the cubes into groups of 100s (10 stacks of 10 cubes each), 10s and 1s.

Teacher Tips

If you do both of the above activities, have the children compare the collections that have been formed and discuss ways of determining the difference between your age and their combined ages. The two collections can also be combined to show the total of all the ages in the room.

Additional Ideas

3 Getting to Know You—Probability

You Will Need
- Chapter 9, Probability
- 0–5 number cube
- several slips of 1 × 4 paper
- a brown bag or container to hold the slips of paper

Your Lesson

(1 day) Introduce basic concepts of probability to your students by having them engage in experiments that involve chance. Today's experiment is an enjoyable way for them to begin thinking about chance and to learn about each other at the same time. Follow these steps:

Each child rolls the number cube once. The children sort themselves into teams according to the numbers rolled. As an example, Team 4 is made up of those children who rolled the number 4.

The team number determines how many slips of paper each team receives. Thus, Team 4 is given 4 slips, Team 2 takes 2 slips, etc. The teams then write their number on each of their slips.

All slips of paper are collected in a container and mixed thoroughly.

Tell the class you will draw a slip of paper from the container 20 times. After each draw, the number on the slip will be tallied and the slip will then be put back into the container.

Ask the children to predict which team number is likely to be drawn the most and to explain their thinking. Why do they feel their predicted results will occur?

Conduct the experiment. Periodically stop to discuss the information that has been gathered. Does anyone want to change their prediction?

Compare the final results with the children's predictions: What explains any agreement or disagreement? If the experiment were conducted again, would the results be the same? Is it possible that Team 1's number will be drawn the most? Is it likely? Why? Was this experiment fair to all teams? Why? If it isn't fair, how could it be changed to remove the unfairness?

Teacher Tips

Theoretically, if the draws are made in a completely random fashion, the number that is on the most slips will have the highest probability of being selected most often. However, this may not happen due to chance (or non-random drawing).

Journal Writing

The children can summarize the results of this experiment and describe their feelings about it. How did the results compare with their predictions? What might explain any disagreement?

4 Getting to Know You—Money

You Will Need
- local grocery advertisements
- calculators

Your Lesson

(1 day) You and your children have enjoyed getting to know one another and now it's time to celebrate the new friendships!

The children will enjoy planning a Getting To Know You party. What refreshments would they like? What other items (e.g., cups and napkins) will be needed?

Check the grocery ads for prices. Will $10 be enough to fund the party? How much money would each child need to bring if the affair is Dutch treat?

Teacher Tips

This activity motivates the use of estimation and mental arithmetic and helps the children work with money in an enjoyable way. Exact totals can also be determined with the help of calculators.

Another lesson that will achieve similar purposes is to use Back-to-School advertisements to determine the cost of school supplies.

Additional Ideas

5 Place Value—Base Five Pieces

You Will Need
- Chapter 4, Place Value; Chapter 3, Patterns
- base five counting pieces for each child (Blackline 5)
- overhead base five counting pieces

Your Lesson

(1 day) Tell the children the names of the base five pieces. Lead your children in a show-and-tell discussion of these pieces and of the relationships and patterns that exist among them. (Please refer to pages 23 and 24 of Chapter 4.)

Emphasis

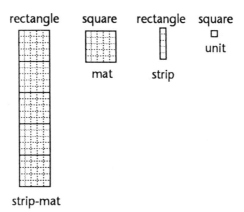

Each piece has 5 times as many units as the piece to its right.

When the different pieces are arranged in order from smallest to largest, they form a two-step, growing pattern of square, rectangle, square, rectangle, etc.

Teacher Tips

The children may make the connection between units, strips and mats with pennies, nickels and quarters, respectively. If so, *great*!

You can extend this lesson by having children predict the size and shape of the next piece in the collection. They will enjoy arranging 5 strip-mats together to form a larger square or "mat-mat".

Additional Ideas

6 Place Value—Totaling Units Using Base Five Pieces

You Will Need
- Chapter 4, Place Value
- base five counting pieces for each child
- overhead base five counting pieces

Your Lesson

(2–3 days) After a brief review, ask your children to set out the following collection of base five pieces:

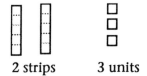

2 strips   3 units

Pose the following question: How many unit squares are in this collection altogether? (13) Have children share, at the overhead, their methods for answering this.

Some other questions to consider:

How many units are represented by the 2 strips? (10) How many units are represented by the 3 units? (3) How many pieces are used in this collection? (5)

Repeat the above activity using different collections of pieces.

Emphasis

Encourage the children to look for different ways to calculate the total number of units in each collection.

For each collection discussed, ask the children to identify the number of pieces used and the number of units represented by the different pieces within the collection.

6 Place Value—Totaling Units Using...(continued)

Teacher Tips

Just to get your juices flowing, we have included more examples and answers in the illustration below. However, do create examples of your own and keep your children on their toes by varying the order of the questions.

strip-mats	mats	strips	units	number of pieces	units represented by strip-mats	units represented by mats	units represented by strips	number of individual units	total number of unit squares
	1	3	2	6	0	25	15	2	42
1	2	0	3	6	125	50	0	3	178
	4	3	4	11	0	100	15	4	119
	1	6	6	13	0	25	30	6	61
1	0	8	0	9	125	0	40	0	165
	2	2	1	5	0	50	10	1	61

Additional Ideas

7 Place Value—Comparing Collections

You Will Need
- Chapter 4, Place Value
- Transparency of Blackline 96 (More or Less spinner)
- a 0-1-2-3-4-3 number cube
- a mats-strips-units cube
- yarn necklaces (see Materials Guide for distribution instructions)
- base five counting pieces for each child

Your Lesson

(1–2 days) To provide practice in comparing the number of units in two collections of counting pieces, play the base five version of Mine or Yours.

Divide your class into two teams by distributing the yarn necklaces.

Both cubes are rolled to determine the collection of base five pieces each Team A (yellow) member is to place in their working space. Roll cubes again for Team B (green). The More or Less spinner is used to decide which team is awarded a point.

TEACHER *The roll for the yellow team is—4 strips. Okay, yellow team, place 4 strips in your working spaces.*

CHILDREN *Four strips—that means we have a total of 20 unit squares represented.*

TEACHER *Now for the green team. Your roll is 3 mats.*

CHILDREN *Three mats—we have a total of 75 unit squares represented.*

TEACHER *Let's compare the collections for "more" and "less" unit squares represented.*

CHILDREN *Green has more and yellow less.*

TEACHER *The spin of the spinner will name the winner! Around it goes. Less! Yellow team earns one point!*

A tally mark is recorded on the chalkboard for the yellow team.

Green	Yellow
	ǀ

7 Place Value—Comparing Collections (continued)

Additional rounds are played, as time permits. At the close of the session, the tallies are compared. The final spin of the More or Less spinner determines the winner for the day.

 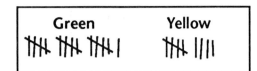

Teacher Tips

You can teach this game to your children by playing against them yourself. You are one team and the entire class is the other team. You can use the overhead base five pieces, while your children use their individual collections.

Additional Ideas

8 Place Value—Representing a Number by Different Collections

You Will Need
- Chapter 4, Place Value
- overhead base five counting pieces
- a wallchart like the one pictured below

Your Lesson

(1–2 days) In this lesson, children represent a given amount of units by different collections of base five pieces and minimal collections are introduced.

Begin with a question such as: How can a total of 17 units be shown with the base five pieces? Challenge your children to show-and-tell different ways of answering this question.

Make a wallchart listing the various collections found, identifying the number of pieces used in each case.

Strips	Units	Number of Units
0	17	17
1	12	13
2	7	9
3	2	5*

Which collection uses the fewest number of pieces to total 17 units? This is known as the *minimal collection* (marked with an * above).

Repeat the above activity beginning with other total amounts of units. Here are some questions to discuss with the children:

How many different collections can be shown for a given amount of units? How many pieces are used in each collection? Which collection uses the least number of pieces?

Why would the least number of pieces be preferred on occasion?

8 Place Value—Representing a Number...(continued)

Emphasis

A given amount of units can be represented in several ways.

Of the different ways to represent a number of units, the one that uses the fewest pieces is called the *minimal collection*. For example, the minimal collections for 17 and 168 are:

    17   3 strips and 2 units
    168   1 strip-mat, 1 mat, 3 strips and 3 units.

Teacher Tips

Excitement is in the air as your children tell and share. Did you ever think you could show 17 in so many ways?

It helps to record the collections for each number with a different color.

Journal Writing

How do you feel about working with the base five pieces?

Additional Ideas

9 Place Value—Identifying the Minimal Collection

You Will Need
- Chapter 4, Place Value
- overhead base five counting pieces
- a wallchart like the one used in Lesson 8

for each child
- base five counting pieces
- a copy of Blackline 39 (Mats, Strips, Units)

Your Lesson

(1 day) In Lesson 8, children formed different collections of base five pieces that represented a given amount of units. This lesson is a variation of the same theme, only this time they begin with a collection of pieces and make trades to determine the minimal collection.

Place a collection of 3 strips and 7 units on the overhead. Record information about this collection on the first line of a chart as shown below. Exchange one of the strips for 5 units and enter the information about the new collection on the chart.

Strips	Units	Number of Pieces
3	7	10
2	12	14

Ask children to record these results on Blackline 39 and add to them by making further exchanges of the pieces. Discuss their results and compile a complete list of the possible collections.

As time permits, repeat this activity beginning with other collections of pieces. For each example, discuss the following questions: What kinds of trades are possible? How many pieces are used in each collection? What is the minimal collection and how can it be formed? What is the total number of unit squares in each collection?

Emphasis

Each collection has the same value. That is, if each collection were exchanged for units, all would contain the same number of units. In the example given above, the number of units is 22.

9 Place Value—Representing a Number...(continued)

The minimal collection will never require more than 4 of any one piece, because any group of 5 or more of the same number piece can be traded for the next larger size. Here is a list of the collections possible for the given example. The minimal collection is marked with an asterisk.

Strips	Units	Number of Pieces
3	7	10
2	12	14
4	2	6*
1	17	18
0	22	22

Journal Writing What makes the minimal collection unique?

Additional Ideas

10 Place Value— Forming Minimal Collections

You Will Need
- Chapter 4, Place Value
- base five counting pieces for the overhead
- calculators (as needed)

for each child
- base five counting pieces
- individual chalkboards, chalk and erasers

Your Lesson

(1–2 days) Ask your children to:

Set out the minimal collection for 33 units on their chalkboards.

Show-and-tell their methods for doing this.

Sort the pieces of their collection by size and arrange them in order from largest to smallest as shown below.

Record the number of times each different piece is used in the collection.

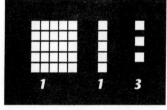

Provide further independent practice by having the children repeat the above activity for other amounts of units. Examples are given below, each with their corresponding minimal collection (this might be an appropriate time to make calculators available to your children). Note that a 0 should be written to indicate the absence of a particular base five piece in a collection.

136	1 strip-mat, 0 mats, 2 strips, 1 unit
98	3 mats, 4 strips, 3 units
74	2 mats, 4 strips, 4 units
190	1 strip-mat, 2 mats, 3 strips, 0 units
144	1 strip-mat, 0 mats, 3 strips, 4 units
103	4 mats, 0 strips, 3 units
6	6 units

10 Place Value—Forming Minimal...(continued)

Emphasis

The recorded numerals indicate the number of times each different base five piece is used in the minimal collection.

A 0 indicates the absence of a particular piece in a collection; in this case, the different pieces within the collection will not alternate between squares and rectangles.

Teacher Tips

We find it great fun to observe the different methods the children use to form minimal collections. This is a wonderful opportunity to appreciate their problem-solving skills.

Journal Writing

How would you describe the minimal collection for a number? What procedure would you use to form the minimal collection for 37?

Assessment

(Optional) Use a +/– checklist to assess children as they work. As you circulate, note if they are: creating the required minimal collection; recording the number of times each different piece appears in the collection; aware of the total number of units represented by the collection.

Additional Ideas

11 Place Value—
Forming Minimal Collections

You Will Need
- Chapter 4, Place Value
- a 0-1-2-3-4-3 number cube
- base five counting pieces for each child
- yarn necklaces (optional)

Your Lesson

(1–2 days) To practice trading and forming minimal collections of base five pieces, play Up and Back Again.

Divide class into two teams. (We like to use the yarn necklaces for this purpose.)

Alternately roll the number cube for each team to determine the number of units that are added to the team's collection. (Each team begins with 0 units.)

As this is done, the team creates the minimal collection of pieces by trading.

The object of the game is for each team to reach an exact total of 25 units (1 mat, 0 strips, 0 units) and reverse to reach an exact total of 0 units. The first team to do so wins.

Teacher Tips

As in Lesson 7, you could teach this game to the children by playing against the class. They love the challenge of trying to beat the teacher! It's a *natural*! (Note: If you play this way, you will need base five pieces for the overhead.)

Additional Ideas

12 Place Value—Addition and Subtraction

You Will Need
- Chapter 4, Place Value
- base five counting pieces for each child
- base five counting pieces for the overhead

Your Lesson

(1–2 days) Challenge your children with problems such as:

Combine 2 strips and 2 units with 1 strip and 4 units. How many units are there altogether? Represent this number of units by its minimal collection.

Start with 4 strips and 3 units. Now remove 1 strip and 4 units. How many units are left?

Have the children show-and-tell their methods. Refer to the dialogue on pages 25 and 26 of Chapter 4 for more information.

Emphasis

What reasoning did the children follow to obtain their answers?

What methods can be used to form a minimal collection?

Encourage the children to share their use of different methods for solving the same problem. Continue to ask the children to explain *how* an answer was obtained.

Teacher Tips

Your children may pose problems of their own. It helps to mix up addition and subtraction examples and to use collections involving a variety of pieces.

Journal Writing

Describe how collections of pieces can be combined.

Challenge

What coins follow the base five counting pattern for three places? (penny, nickel, quarter) Challenge the children to create a "coin" or "bill" for $1.25, which would be equivalent to a strip-mat. How would it look? From what material would it be made? Get the creative juices flowing. Share the ideas! They might make an interesting bulletin board display.

Additional Ideas

13 Place Value—Making Base Two Pieces

You Will Need
- Chapter 4, Place Value; Chapter 3, Patterns
- several sheets of 2-cm grid paper (Blackline 73)
- scissors
- paper (for background), glue, crayons

Your Lesson

(1 day) Ask children to imagine base two counting pieces. Invite volunteers to describe the different pieces and the number of units they contain.

Now have the children cut the individual base two pieces from grid paper and color and display them. Such a display might look like:

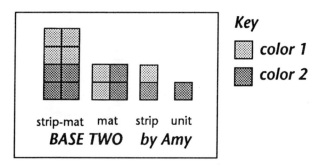

Discuss the following questions: What patterns are present in the base two pieces? How are the base five and base two pieces similar? different?

Emphasis

Base two and base five pieces are alike in structure except that each base two piece represents twice as many unit squares as the next smallest one (that is, each piece doubles in size). Beginning with the unit, the different pieces in both sets will form an alternating square-rectangle pattern.

13 Place Value—Making...(continued)

Teacher Tips

Since it will likely take an hour to complete this lesson, you might omit the Contact lesson on this day.

You can skip the cutting part of this lesson and have the children simply color the different pieces on 2-cm grid paper. In either case, our children have found it helpful to show the relationship between the pieces by using two different colors. For example, a mat can be shaded in a way that shows the two strips that make it up.

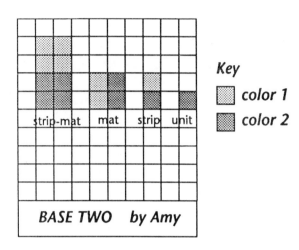

Keep the display for use in the following lesson.

Additional Ideas

14 Place Value— Making Base Three Pieces

You Will Need
- Chapter 4, Place Value; Chapter 3, Patterns
- several sheets of 2-cm grid paper (Blackline 73)
- scissors, glue, tape, markers or crayons for each child
- the base two displays from Lesson 13
- paper for background

Your Lesson

(1 day) Follow the procedures for Lesson 13 and have your children prepare a display of base three counting pieces.

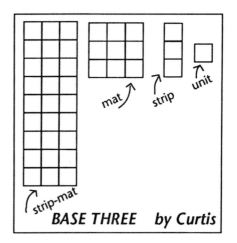

What patterns are present in the base three pieces?

How are the base five, base two and base three counting pieces alike? different?

Emphasis

Each base three piece represents 3 times as many unit squares as the next smallest one. Thus, beginning with the unit, each piece triples in size.

Regardless of the base, if the different pieces are arranged in order from smallest to largest, they will form an alternating square-rectangle pattern.

14 Place Value—Making Base Three...(continued)

Teacher Tips

An alternative to a cut-and-paste display is to color the pieces on 2-cm grid paper as shown here:

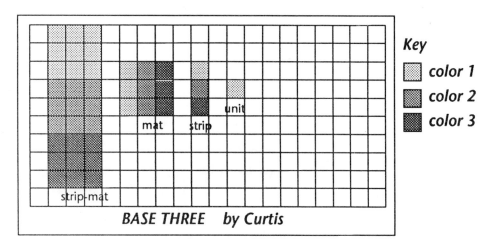

Since this lesson may take an hour to complete, you may wish to skip the Contact lesson for this day.

Your children might enjoy extending this activity by making a set of base four counting pieces at home.

Our first open house was quite rewarding when our children explained these displays to their parents!

Journal Writing

How would you describe the pattern of the base three pieces? How are they like base five pieces? How are they different?

Additional Ideas

15 Place Value—Making Base Eight Pieces (Optional)

You Will Need
- Chapter 4, Place Value; Chapter 3, Patterns
- several sheets of 2-cm grid paper (Blackline 73)
- scissors
- individual copies of Blackline 39 (Mats, Strips, Units)

Your Lesson

(2 days)

Day 1

Ask the class to imagine base eight counting pieces and to describe the different pieces.

In pairs, have children cut a collection of these pieces from grid paper. Each team's collection should consist of 1 mat, 7 strips and 7 units.

Discuss these pieces:

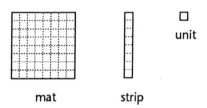

What patterns can be observed? What will a strip-mat look like?

Day 2

Ask each team to form a collection of 4 strips and 4 units from its set of base eight pieces. Challenge them to calculate the number of unit squares represented by this collection (Answer: 36 units). Each team member records the information on Blackline 39. Share methods used to calculate this total.

15 Place Value—Making Base Eight Pieces (continued)

Continue in the same manner with examples such as those below. Have the children calculate the total number of unit squares in each collection.

Mats	Strips	Units	Total Unit Squares
1	5	3	107
0	6	5	53
1	7	7	127
1	2	3	83
0	7	1	57
0	5	5	45
0	6	0	48
1	1	1	73
1	0	6	70

Have children share their thinking as they determined the total number of unit squares represented by the pieces.

Emphasis Each base eight piece represents eight times as many unit squares as the next smallest one. A strip-mat will be a rectangle that is formed by arranging eight mats in a line and taping them together.

Teacher Tips You guessed it! You are well on your way to base ten!

Homework (Optional) Challenge your children to create a set of base six pieces from grid paper.

Additional Ideas

16 Place Value—Base Ten Awareness

You Will Need
- Chapter 4, Place Value
- base ten counting pieces for each child
- overhead base ten counting pieces

Your Lesson

(1 day) Your children are now ready to be introduced to the base ten counting pieces. Begin by asking them to close their eyes and imagine a base ten unit, strip and mat. While keeping their eyes closed, have them describe the pieces and the number of units each piece contains. Put these pieces on the overhead and have the children open their eyes. How do their mental pictures compare to what they now see?

Your children will enjoy having a short time to play freely with the pieces. They then may be interested in building a strip-mat or in thinking about what a mat-mat might look like.

As time permits, ask the class to determine the total number of units represented by a collection of 2 mats, 3 strips and 7 units. Do the same with other collections, remembering to include some that represent totals that have 3 digits or require a 0.

Some questions for discussion: How are the base ten counting pieces formed? How many units are contained in the individual pieces? How are these pieces similar to the counting pieces used in previous lessons? How are they different?

How does a collection of 3 strips and 2 units using base ten pieces compare with a collection of 3 strips and 2 units of base five pieces? Which collection represents the greater amount of units? What if one compared like collections of base five and base three pieces?

16 Place Value—Base Ten Awareness...(continued)

Emphasis

Base ten pieces exhibit the same growing pattern as those used in previous lessons, except that each piece represents 10 times as many units as the next smallest piece. Each piece, therefore, is 10 times larger than the next smallest piece. Also, if they are arranged in order, from smallest to largest, they form an alternating square-rectangle pattern.

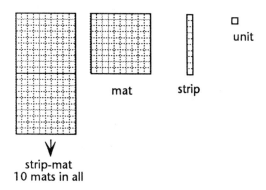

Each strip contains 10 units and each mat contains 10 strips or 100 units. A strip-mat will be a rectangle that contains 1,000 unit squares. It is formed by arranging 10 mats side-by-side in a line and connecting them together. A mat-mat will be a square that is formed by 10 strip-mats. It will contain 10,000 unit squares.

When comparing like collections from two different bases, the collection from the larger base will contain the greater amount of units.

Teacher Tips

This lesson usually takes one day to complete. Because of their experiences in the earlier lessons, the children will find the transition to base ten a very natural and enjoyable one. In fact, after this brief introduction, they are likely to be "champin' at the bit" to do greater things with the pieces. Isn't it wonderful to see understanding building daily?

Journal Writing

Describe the base ten pieces. How are they like the other pieces that have been used? How are they different?

Additional Ideas

17 Place Value—Forming Collections

You Will Need
- Chapter 4, Place Value; Insight Lesson 7 (Mine or Yours)
- a mats-strips-units cube
- one of these cubes: 0–5, 4–9 or 3–8
- yarn necklaces
- transparency of Blackline 96 (More or Less spinner)
- base ten counting pieces for each child

Your Lesson

(1–2 days) To provide practice in comparing the number of units in two collections of base ten pieces, have the children play the base ten version of Mine or Yours.

Divide your class into two teams by distributing the yarn necklaces.

In turn, each team rolls the mats-strips-units cube along with a number cube of its choice. This determines the collection of base ten pieces each team member places in their working space.

The More or Less spinner is used to determine which team is awarded a point for the round. Keep a tally of team points on the chalkboard.

Additional rounds are played as time permits. At the close of the session, the point totals for each team are compared. The final spin of the More or Less spinner determines the winner for the day.

Additional Ideas

18 Place Value—
Ways to Represent a Number

You Will Need
- Chapter 4, Place Value
- overhead base ten counting pieces
- transparency of Blackline 39 (Mats, Strips, Units)

for each child
- base ten counting pieces
- a copy of Blackline 39 (Mats, Strips, Units)

Your Lesson

(2–3 days) How can 21 units be shown with the base ten pieces? Challenge your children to find different ways of answering this question by making equal exchanges of the pieces, such as 1 strip for 10 units. Have them then show-and-tell their work, recording information on Blackline 39 about each collection found.

Strips	Units	Total Number of Pieces
0	21	21
1	11	12
2	1	3*

The minimal collection (marked above with an asterisk) should be identified.

Repeat the above activity using other amounts of units. Have the children look for the minimal collection in each case. As another example, the chart below shows the collections that represent 42 units (see Chapter 4, pages 26 and 27 for more examples). Vary the size of the numbers to include some with 3 digits. Provide time for children to make observations about this work. In particular, do they notice any connection between the total number of units that is being represented and its minimal collection?

Strips	Units	Total Number of Pieces
0	42	42
1	32	33
2	22	24
3	12	15
4	2	6*

18 Place Value—Ways to Represent...(continued)

Emphasis

A given amount of units can be represented in several ways with the base ten pieces. A mat can be exchanged for 10 strips and a strip for 10 units.

The minimal collection needed to show a total number of units can be identified from the number itself. For example, here are the minimal collections for totals of 21 and 102 units:

21 units   2 strips and 1 unit.

102 units   1 mat, 0 strips and 2 units.

This connection provides your children with a visual model for thinking about place value and positional notation. Notice that a 0 indicates the absence of a piece in a collection.

With the base ten pieces, a minimal collection for a number will never require more than 9 of any one piece because any group of 10 or more of the same piece can be traded for the next larger size.

In this program, numbers are most often represented by minimal collections of base ten pieces. However, non-minimal collections are sometimes used when performing computations.

Teacher Tips

The above activities will help the children grow in their understanding of place value and at this point they will often quickly see the role played by the minimal collection.

Be sure to include examples that make use of 0s.

Our children regard the minimal collection as the "easiest collection" to carry around. What joy it has given us to hear them exclaim, "I see the pattern! This is easy!"

Journal Writing

Describe how you would proceed to form the minimal collection for a total of 107 units. What does the 0 in 107 mean? How do you feel about using base ten pieces?

Assessment

(Optional) A "+ or –" checklist and a review of journals would work nicely here.

Additional Ideas

19 Place Value—
Using Base Ten Grid Paper

You Will Need
- Chapter 4, Place Value
- a Number Card packet for every four children
- base ten counting pieces for the overhead
- transparency of Blackline 8

for each child
- several sheets of base ten grid paper (Blackline 8)
- crayons or markers for making diagrams
- base ten counting pieces

Your Lesson

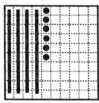

4 strips and 6 units
46

(1 day) Introduce your class to base ten grid paper. Discuss what it is and why some lines are darker than others. Illustrate how the paper can be used to portray mats, strips and units.

Have your children now form the minimal collection of pieces needed to represent a total of 46 units. Be sure to review how the answer of 4 strips and 6 units can be identified from the number 46. Complete the discussion by asking the children to diagram this answer on base ten grid paper.

Challenge the children to form the minimal collections for other numbers and to diagram their answers on grid paper. Include examples that use mats as well as strips and units. Encourage them to work at their own pace, completing as many as possible. After a suitable time, ask people to share some of their diagrams at the overhead. Here are three more examples:

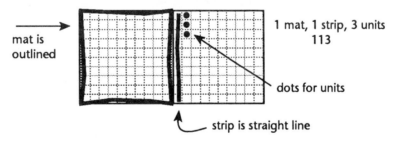

1 mat, 1 strip, 3 units
113

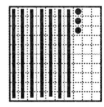

7 strips and 3 units
73

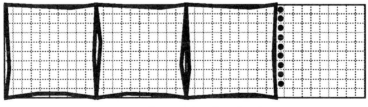

3 mats, 0 strips, 9 units
309

19 Place Value—Using Base Ten...(continued)

Discuss the minimal collection for each number you use. How can it be spotted by looking at the number itself?

For more practice, give each group of four a packet of Number Cards. Each child can then draw numbers from the packet and diagram them.

Emphasis

There are at least two options for displaying a number of units—a physical model or a diagram. Both are helpful for building a "mental picture" of the number.

Grid paper is lined to make diagramming easier.

Colored diagrams are more easily seen than lead pencil ones. It is not necessary to be overly meticulous when diagramming, however.

Teacher Tips

Allow each child to work at a comfortable pace. Some children are going to complete more samples than other, just as some children run faster than others. Wise use of time and remaining on task are much more important than speed or amount of work completed. Adjust the pace to meet the needs of your class, repeating the lesson as needed.

Homework

(Optional) For the next several days, assign homework that calls for your children to diagram various numbers on grid paper. The numbers can come from grocery ads, football scores, uniform numbers of players, measures in a musical piece, temperatures observed during the day, prices of toys, etc. We have also found that the assignment becomes more fun if the children generate sources for the numbers to be used.

Additional Ideas

20 Place Value—Drawings

You Will Need
- Chapter 4, Place Value
- base ten counting pieces
- base ten grid paper (Blackline 9)
- crayons or markers
- Number Card packet for every three or four children

Your Lesson

(2–3 days) Conduct a show-and-tell discussion of the different ways to represent 48 units with base ten pieces. Note the minimal collection of 4 strips and 8 units.

Follow this review by asking your children to form some of the collections that show 56 units with their base ten pieces. Have them diagram and label each collection on grid paper, indicating the minimal one. Some possibilities are shown here:

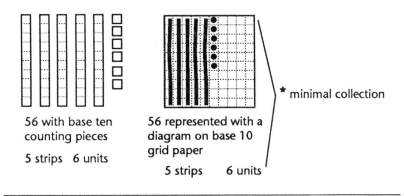

56 with base ten counting pieces

5 strips 6 units

56 represented with a diagram on base 10 grid paper

5 strips 6 units

* minimal collection

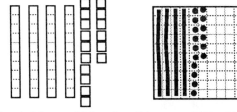

56 = 4 strips and 16 units

Diagram of 56 with 4 strips and 16 units

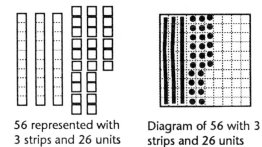

56 represented with 3 strips and 26 units

Diagram of 56 with 3 strips and 26 units

20 Place Value—Drawings (continued)

As the children finish, conduct a show-and-tell discussion of their diagrams at the overhead. Have them also compare their diagrams with others in the room.

Provide independent practice for this activity by using numbers drawn from the Number Cards. Here is an example showing a collection of 123 units.

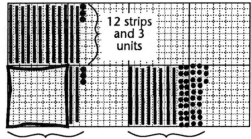

1 mat, 2 strips and 3 units   8 strips and 43 units
*the minimal collection

Emphasis

A given number of units can be diagrammed in many ways. Each way corresponds to a specific collection of base ten pieces.

It is not necessary that all of the collections showing a given number be formed. The main point is that, while one most often thinks of numbers in terms of minimal collections, there are times when it is helpful to visualize other representations. As an example, one can picture 123 as 1 mat, 2 strips and 3 units *or* as 12 strips and 3 units. This is analogous to thinking of $1.23 as 12 dimes plus 3 pennies.

Teacher Tips

To link this lesson with some of the Contact activities, ask the children to estimate the number of apple-flavored cereal pieces in a container. Their estimates can be diagrammed in the manner described above.

Additional Ideas

21 Place Value—Expanded Forms

You Will Need
- Chapter 4, Place Value
- various forms of base ten grid paper (Blacklines 8–13), as needed, and crayons or markers
- transparencies of Blacklines 8–13 and marking pens
- packets of Number Cards for every four children

Your Lesson

(2–3 days) Have a volunteer diagram the minimal collection for 135 units at the overhead. Discuss the number of units represented by the different pieces in the diagram and indicate them as shown here:

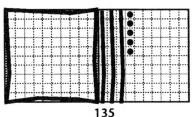

135
1 mat,   3 strips and 5 units
1 hundred + 3 tens + 5 ones
100 + 30 + 5

The expressions 1 hundred + 3 tens + 5 units and 100 + 30 + 5 are two ways of writing 135 in *expanded form*.

Ask your children to diagram the minimal collection for 242 units on their grid paper and to write it in expanded form.

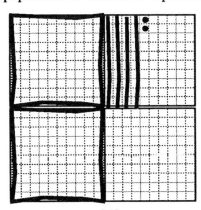

242
2 mats,   4 strips and 2 units
2 hundreds + 4 tens + 2 ones
200 + 40 + 2

149
INSIGHT LESSONS

21 Place Value—Expanded Forms (continued)

Repeat this activity for other numbers, indicating expanded forms for each. Be sure to use some numbers that have a 0 in them. Here are two more examples:

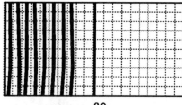

80
8 strips and 0 units
8 tens + 0 ones
80 + 0

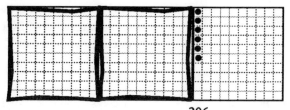

206
2 mats,   0 strips and 6 units
2 hundreds + 0 tens + 6 ones
200 + 6

For additional practice, use the Number Cards as a source of numbers for children to diagram and write in expanded forms.

If a minimal collection is diagrammed, how many units are represented by each of the different pieces shown? How can a number be visualized in expanded form?

Emphasis

Expanded forms of a number are helpful because they provide information about the amount of units represented by the different base ten pieces in the minimal collection. They also highlight the place value of each digit in a number.

Relating the expanded form of a number to the use of base ten pieces (and diagrams) will strengthen your children's visual perception of place value.

Teacher Tips

This type of activity is important and you may wish to spend a bit more time with it at this point. In any case, we encourage you to refer to expanded forms when conducting other number lessons throughout the year.

Journal Writing

Complete this sentence: Today I learned that _____.

Homework

(Optional) This is an opportunity to involve family members in the search for information to diagram. Have your children sketch and indicate the expanded form for the number of days in a year, the cost of a meal at a fast food restaurant, the number of ounces in a gallon of milk, the length of a favorite television show in minutes, the number of classified ads on a page in the newspaper, etc.

22 Place Value—More and Less

You Will Need
- Chapter 4, Place Value
- base ten counting pieces for each child
- base ten counting pieces for the overhead

Your Lesson

(1–2 days) In this lesson, your children will form minimal collections of numbers that are arranged in order. In so doing, they will gain further experience with making equal exchanges of base ten pieces and learn about the relative sizes of numbers. The main activity of the lesson is described in the following example:

Have children form the minimal collections for these three numbers: 35, 2 less than 35 and 2 more than 35. Identify each of the numbers represented.

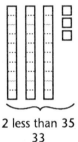 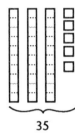 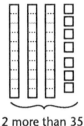

2 less than 35   35   2 more than 35
33   37

Continue by asking the children to form minimal collections for other sets of numbers and conduct a show-and-tell discussion of their work. Here are some examples you might use:

46, 10 less than 46 and 10 more than 46

39, 3 less than 39 and 3 more than 39

42, 5 less than 42 and 5 more than 42.

108, 13 less than 108 and 13 more than 108

230, 28 less than 230 and 28 more than 230.

Teacher Tips

Please refer to Chapter 4, pages 28 and 29, for additional discussion of this activity. In particular, note that it is likely that your children will trade the number pieces in different ways as they pursue these problems. Provide numerous opportunities for them to share their thinking. We have found that this activity reinforces place value concepts and, at the same time, prepares children for later work with number progressions and with addition and subtraction.

Additional Ideas

23 Place Value— More-and-Less Diagrams

You Will Need
- Chapter 4, Place Value
- various forms of base ten grid paper (Blacklines 8–13) and crayons
- transparencies of Blacklines 8–13

Your Lesson

(1–2 days) Conduct this lesson just as you did Lesson 22, substituting base ten grid paper diagrams for the physical model.

Teacher Tips

Some children will enjoy the challenge of extending the number progressions. For example, they might diagram the following numbers: 35, 2 less than 35, then 2 more, etc.

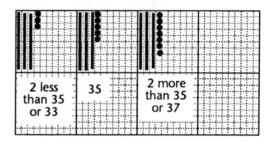

Homework

(Optional) More-and-less diagrams can provide opportunities for some family entertainment. Some fun assignments to diagram: the age of the person with the longest hair, then 2 more and 2 less; the number of light bulbs in the house, then 6 more and 6 less; the amount of change in a wallet, then 11 more and 11 less; the combined heights of family members in inches, then 25 more and 25 less; the combined letters in your name, then 4 more and 4 less.

Additional Ideas

24 Place Value—Comparing Numbers

You Will Need
- Chapter 4, Place Value
- various forms of base ten grid paper (Blacklines 8–13) and crayons
- transparencies of Blacklines 8–13
- Number Card packets for every two children

Your Lesson

(1–2 days) Conduct a discussion of the symbols <, > and =. Tell your children that someone long ago chose to use < to stand for the phrase "is less than". Brainstorm possible reasons for this choice.

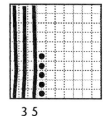

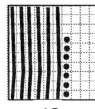

35 < 67

Have your children diagram 35 and 67 and indicate how these numbers are related with the appropriate symbol. Ask volunteers to describe the thinking they used to compare the numbers.

Examine other pairs of numbers in the same way. You might select from the Number Cards or present examples such as:

The number of states in the United States and the number of senators in Congress.

The number of boys and the number of girls in your class.

The number of stripes on the American flag and the number of states touching the Mississippi River.

50 < 100
number of states < number of senators

Journal Writing

Why do you think the symbol > was chosen to represent "greater than"?

Teacher Tips

We don't know why the symbols for "less than" and "greater than" were chosen. The symbols were first used by Thomas Harriott in the 17th century. Our children have enjoyed speculating about why these symbols might have been selected and this led to a nice creative writing assignment.

Homework

(Optional) Ask the children to diagram and compare some more of the numbers in their world, such as: the number of presidents we have had and the number of senators in Congress; the number of steps in the Statue of Liberty and those in the Empire State Building; the longest estimated length of Brontosaurus and that of Tyrannosaurus Rex; the number of countries north and south of the Tropic of Cancer.

25 Place Value—Up and Back Again

You Will Need

Versions 1 and 2
- two number cubes, 0–5 and 4–9

Version 3
- three number cubes, 00–50, 0–5 and 4–9
- base ten counting pieces for each child

Your Lesson

(2–3 days) To practice trading and forming minimal collections of base ten pieces, play the following versions of Up and Back Again.

Version 1

Establish two opponents such as:

a. teacher vs. class.

b. half the class vs. the other half (yarn necklaces could identify each team).

c. child vs. child. (In this case, each pair of children will need number cubes. The small wooden cubes from MLC work nicely. Numbers can be written on them with a permanent marker.)

The teams alternately select one of the number cubes to be rolled. Each roll determines the number of units to add to that team's collection.

After adding the pieces, the team is to create the minimal collection of pieces by appropriate trading.

The object of the game is to reach an exact total of 50 units and then reverse to reach 0 units (exactly). The first team to do so wins.

Version 2

Play as above, however, in this version, each team may choose to roll *one or both* cubes. If both are tossed, the sum of the numbers rolled determines the amount of units added to the team's collection. The object is to reach an exact total of 100 (or whatever total the children predetermine) and then reverse to reach an exact total of 0 units. The first team to do so wins.

Version 3

For this version, you will need the following number cubes 00–50, 0–5 and 4–9. Play as above, however this time each team may choose to roll either *one, two or three* cubes. Again, the total of the rolled numbers determines the amount of units added to a team's collection. The object is to reach an exact total of 300 units (or whatever total the children predetermine) and then reverse to reach an exact total of 0 units. The first team to do so wins.

Additional Ideas

26 Communication of Process

You Will Need
- base ten counting pieces for each child
- base ten counting pieces for the overhead

Your Lesson

How many times have you driven home from work and upon arrival find yourself unable to remember any specifics of your journey? Too often children journey through "manipulative math", moving objects without focusing on the process (the sequence of and the reasoning behind their actions).

This activity encourages children to state their process clearly and completely. By sharing it with a friend(s), they are forced to think through their method to explain and justify it.

There are three versions of this activity, each with a different focus:

Version 1

Focus: Auditory commands and verbal communication of process.

Teacher calls out:	Children:
34	Form a collection for 34. (May or may not be the minimal collection.)
	Share their methods for creating a collection for 34 (either at the overhead or in small groups).
65	Add pieces to their existing collection for 34 to form a collection for 65.
	Are encouraged to "think aloud" as they move their pieces.
	There are several possibilities.
45	Adjust pieces to create a collection of 45.
	Share their thinking aloud.
27	Adjust pieces to create a collection of 27.
	Share their thinking aloud.

Version 2

Focus: Visual commands and auditory communication of process.

In this version you will be writing the numbers rather than calling them out (as you did in Version 1). In this activity ask the children to always set up the minimal collection for the given number.

Write 231 in a place clearly visible to all.

The children look at it and set up the minimal collection, consisting of 2 mats, 3 strips and 1 unit. Each child shares with a friend what pieces they chose and why they chose these pieces.

26 Communication of Process

Erase the 1 and replace it with a 2, making 232.

The children look at the number and respond by adding 1 more unit. Again a verbal description of process is exchanged among the children. Volunteers may wish to share their process with the whole class in a show-and-tell style.

Change the 3 to a 5, and the children adjust their collection to show 252.

Continue in the same manner, remembering to increase the challenge by changing more than one digit!

Version 3

Focus: Auditory and/or visual commands and written communication of process.

In this version speak or write the numbers and ask the children to communicate their process in a written form that includes words, pictures and/or numbers.

Teacher Tips

You will want to revisit this lesson often.

Having to work through the communication of process, both verbally and on paper, works on long-term memory. By practicing visual thinking (revisiting the problem and action visually) retention is greatly increased.

Your children will need to learn to describe process using pictures only, numbers only and words only. In addition they will need to develop a communication style that uses any combination of pictures, numbers and words.

If your children are having difficulty describing process try "spotlighting" one child's process. Lead your children in outlining or webbing the sequence of the process as demonstrated by the child. Create a class chart of the child's process. Such a modeling of clear, complete descriptions is priceless.

You may wish to gather video clips of your children as they problem-solve with pieces. To help you children learn *how* to write about process, show a clip, pausing occasionally to record bits and pieces on a class chart. Before long, your children will develop the communication and sequencing skills needed to describe their own processes independently and in detail.

27 Place Value—Sketches

You Will Need
- Chapter 4, Place Value
- several sheets of unlined paper for each child
- Number Card packets

Your Lesson

(1–2 days) How would children diagram 46 if grid paper is not available to them? One suggestion is to make a freehand sketch; invite them to try this and to then share their work. Continue by asking them to sketch other numbers (perhaps drawn from the Number Cards), challenging them to show 4-digit numbers as well.

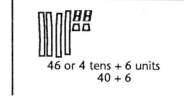

46 or 4 tens + 6 units
40 + 6

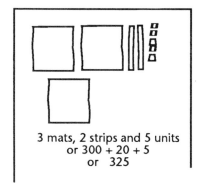

3 mats, 2 strips and 5 units
or 300 + 20 + 5
or 325

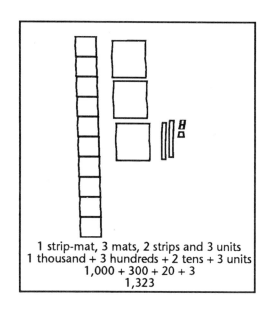

1 strip-mat, 3 mats, 2 strips and 3 units
1 thousand + 3 hundreds + 2 tens + 3 units
1,000 + 300 + 20 + 3
1,323

Near the end of this lesson, ask your children to solve some place value riddles. Here are some examples:

(Hold up a large handful of units.) I have 348 units in my hand. Imagine what the minimal collection would look like—can you describe or sketch it?

My number is 2 strips more than 203. What is it? What does the minimal collection for my number look like?

My number is 1 strip less than 9 strips and 9 units. What is my number?

Imagine 2 mats, 9 strips and 7 units. My number has 5 more units than this. What is my number?

Ask volunteers to discuss their solutions show-and-tell style.

27 Place Value—Sketches (continued)

Emphasis

Sketching provides a fairly quick, personal way for children to portray numbers, including 4-digit ones.

Sketching is a helpful problem-solving tool. It is used throughout the study of mathematics.

Sketches do not need to be detailed.

Teacher Tips

Encourage the use of sketches. They provide a quick way for picturing a problem and of describing mental images. Children enjoy this way of displaying numbers.

Be sure to provide time for the children to extend their understanding of place value by asking them to sketch numbers that use strip-mats (1,000s) or even mat-mats (10,000s).

Journal Writing

Ask children to write some riddles of their own. They might even enjoy publishing a *Riddle Book* for all to enjoy.

Assessment

(Optional) Assign numbers for your children to sketch, including some that are 4 digits or have 0s.

Homework

Have the children:

Write the last 4 digits of the telephone numbers for 5 of their friends and sketch "picture" representations of them. Order the numbers from greatest to least.

Cut 2- and 3-digit numbers from the newspaper, glue them on a sheet of paper from greatest to least value and draw freehand sketches of them.

Cut several 3-digit numbers from the newspaper and glue them on a sheet of paper. Beside each, ask them to make sketches of the number, 1 less than the number and 1 more than the number.

Additional Ideas

28 Addition/Subtraction Operations, Number Sense & Process

You Will Need
- base ten counting pieces for each child
- base ten counting pieces for the overhead

Your Lesson

The following versions of this activity will give your children plenty of practice with removing and adding units, trading counting pieces, developing number sense and communicating process.

This activity is a variation of Lesson 26, Communication of Process. Instead of calling out numbers, however, ask the children to add or remove a certain amount of units from the pieces in front of them. Have the children form the minimal collection of pieces after each change.

Version 1

Focus: Auditory commands, show-and-tell process.

Have each child set out 34 (3 strips and 4 units) on their working space.

Teacher calls out:	Children:
Remove 4 units.	Remove 4 units and leave 3 strips.
	Describe "how" and "why".
Add 100 units.	Add a mat
	Show-and-tell process and justification.
Remove 3 units.	Remove 3 units and leave 1 mat, 2 strips and 7 units.
	Share reasoning.
Remove 90 units.	Use any method they choose to remove 90 units and leave 3 strips and 7 units
Etc.	

Throughout the activity a variety of methods should surface. For example, when asked to remove 90 units your children might:

- exchange the mat for 10 strips and then remove 9 strips.
- remove the mat and give back 1 strip.
- cover up all but 1 strip of their mat with a piece of paper. (In this case, the revealed strip along with the original 2 strips and 7 units represents 3 strips and 7 units.)

28 Addition/Subtraction Operations... (continued)

Allow the children freedom to develop their own personal number sense. Provide opportunities for the methods to become contagious through show-and-tell.

Version 2

Focus: Visual commands, show-and tell process.

In this version the changes are to be indicated in writing. For example, if the children have 156 (1 mat, 5 strips and 6 units), you might give them the written signal +12. Children use the process of their choice to show a total of 12 units added to their existing collection (and, as before, forming the minimal collection for the new number).

Version 3

Focus: Auditory and/or visual commands with written communication of process.

In this version, children communicate process in a written form (words, numbers, pictures). By focusing upon the operation as well as the number of units in the exchange, number sense is developed.

Teacher Tips

This is a key lesson. Therefore, you will want to revisit it often.

If your children are having too much difficulty writing clear, complete sentences about mathematical processes, have them write about sequences or processes that are non-mathematical. For example, invite the class to create a chart that describes the process used when sharpening a pencil, making a peanut butter and jelly sandwich, brushing teeth or tying a shoe. Once children learn to write clearly about everyday processes, they will have less difficulty describing mathematical processes with words.

Before asking children to work independently with Version 3, you might use a volunteer and model how to describe their process with numbers, symbols and words. Such modeling helps children learn how others communicate mathematical processes clearly.

All children know how to show off! When asking children to write about their process, encourage them to use this opportunity to "show off" by including every possible detail!

Encourage the use of illustrations and numbers within the text. It is true that "one picture is worth a thousand words." Just imagine how difficult it would be to understand these *Opening Eyes to Mathematics* guides if we had used words only!

Consider creating a collective chart of "the different ways we solved this problem".

29 Place Value—Out of Bounds

You Will Need
- base ten counting pieces for each child
- three number cubes 0–5, 4–9 and +/–

Your Lesson

(1–2 days) Out of Bounds is a game that will help your children get ready for addition and subtraction. It is played in the following way.

The class is divided into two teams and the children decide on two boundary numbers.

Each team begins play with 5 strips.

The teams then alternately roll the +/– cube and one of the number cubes to determine the number of units to be added (or subtracted) from their collections.

The first team to go Out of Bounds is the winner (or loser, as pre-determined by the children). For example, if the boundaries are 25 and 75, the game ends as soon as one team's collection contains less than 25 or more than 75 units.

Teacher Tips

Our children usually choose the boundaries that differ by 20 to 30 units.

Additional Ideas

30 Place Value—Rounding

You Will Need
- Chapter 4, Place Value
- base ten pieces for the overhead
- transparency of Blackline 114 (Ten Strip Boards) *for each child*
- base ten counting pieces
- ten-strip board

Your Lesson

(1–2 days) Have each child set up the minimal collection for 72 on a ten-strip board. Is this number closer to 70 (7 strips) or 80 (8 strips) in value? Why? Have your children show-and-tell their responses and discuss with them the concept of rounding. In this case, when rounding to the nearest 10 (or strip), 72 is rounded to 70.

Examine other numbers in the same way. As another example, have the children set up the minimal collection for 134 and examine the following questions:

Is 134 closer to 100 (1 mat) or 200 (2 mats) in value? Answer: To the nearest hundred (or mat), 134 is rounded to 100.

Is 134 closer to 130 (13 strips) or to 140 (14 strips)? Answer: To the nearest 10 (or strip), 134 is rounded to 130.

Emphasis

Rounding is a method of approximating numbers. It is usually thought of in terms of a given place value (e.g., "round to the nearest hundred") and is useful in making estimates and mental computations.

Teacher Tips

Have your children decide how to deal with "half-way" numbers. Such a discussion with our children led to a comparison of ages. When they are $8\frac{1}{2}$ years old, they really prefer to be considered 9 instead of 8. So they decided to round the half-way numbers up to the higher number. Little do they know how their thinking will change when they are $29\frac{1}{2}$!

You might spend part of the time sketching numbers in order to "see" how to round them.

Additional Ideas

31 Place Value—Rounding

You Will Need

Version 1
- a copy of Blackline 15 for each child
- two number cubes, 0–5 and 4–9

Version 2
- a copy of Blackline 16 for each child
- three number cubes, 0–3, 4–9 and 3–8

Your Lesson

(1–2 days) Practice rounding and graphing numbers by playing Rounding Off. Begin with Version 1 and move on to Version 2 when you sense that your children are ready for a bigger challenge.

Version 1

Divide the class into two teams.

One team rolls both number cubes and arranges them to create a number. For example, if 3 and 6 are rolled, the numbers 36 or 63 can be formed.

That number is then rounded off to the nearest strip (10) and recorded by each team member in the appropriate column of their Blackline 15.

Repeat this process for the other team.

Play continues with the teams alternating turns until one team has filled a column on its blackline.

Version 2

This is played like Version 1, however 3 number cubes are rolled rather than 2. For example, if 3, 6 and 7 are rolled, they could be arranged to create 367, 376, 673, 637, 763 or 736.

Teacher Tips

A great deal of strategy comes into play when a column appears to be filling up. You will find the children working hard to rearrange the numbers to fit them into the column that is "looking good" for their team.

Additional Ideas

32 Place Value—In the Mind's Eye

You Will Need
- Chapter 4, Place Value
- individual chalkboards, chalk and erasers

Your Lesson

(1–2 days) Are the base ten pieces becoming pictures in the mind's eye? Hopefully the answer is *YES!*

Pose several questions to the children and have them respond with numbers and symbols on their chalkboards. The questions should reflect the place value concepts that have been developed in the previous lessons.

Some examples:

Think of 165. What does its minimal collection look like? Write an expanded form of this number.

Can you visualize 230? How about 207? Which is the larger number?

Think of 75. Imagine removing 7 units from it. Sketch a picture of the number you get.

Imagine 508. How many strips are there in the minimal collection? How many mats? If 508 is rounded to the nearest hundred, what number would you get?

I am thinking of 1 strip of mats, 3 strips and 2 units. How many units are contained in this collection?

Have the children share their answers and demonstrate their thinking at the overhead. Also invite them to make up questions for the class.

Teacher Tips

This lesson summarizes the work on place value and we encourage you to add to the list of questions. If your children are comfortable with this activity, they will be able to think about numbers and perform number operations with the help of visual models.

Journal Writing

How would you describe the number 573 to a friend? What picture of this number would you share? Do the same for the number 1,080.

Additional Ideas

33 Exploring the Difference Between Two Numbers

You Will Need
- Chapter 5, Addition and Subtraction
- base ten pieces and ten-strip board for each child
- base ten pieces for the overhead
- transparency of Blackline 114 (Ten Strip Boards)

Your Lesson

(3–5 days) During Insight Lessons 23 and 24, the children used base ten pieces and diagrams to decide if two numbers were equal or not. In this lesson, they extend that experience by constructing visual methods for determining the difference between two unequal numbers.

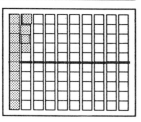

Pose this problem to the class: "Ray and Martha have been collecting pennies. Last week, Ray collected 14 pennies and Martha collected 42. How many more pennies than Ray did Martha collect last week?"

Represent Martha's and Ray's collections on the overhead ten-strip board as shown and ask the children (in pairs) to do the same at their desks using green base ten pieces.

Ask the children to describe how they would estimate an answer to this question. Then, with their partners, have them search for ways to show the exact difference between the two collections. After a period of working time, ask volunteers to show-and-tell their procedures. Here are two possible responses:

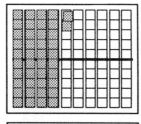

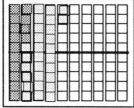

"I added yellow pieces to the 14 to make it look like 42. I needed 2 strips and 8 units. Martha collected 28 more pennies."

33 Exploring the Difference...(continued)

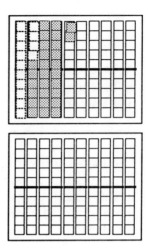

"I covered up part of the 42 with the 14. Twenty-eight units are not covered so Martha has 28 more pennies."

Allow plenty of time for various methods to develop and to be discussed. We usually spend a whole day on this one problem! It is important that the children have an opportunity to construct several ways for showing the difference between 42 and 14 with their pieces and to share their thinking. Challenge them to "find another way," and of course, you can share one of your own!

Over the next few days, continue to ask the children to model "difference" stories with base ten pieces. Encourage estimation and be sure to use some numbers that have 3 digits or 0s. The children will also enjoy making up stories of their own!

Emphasis

The difference model of subtraction is very powerful. Allow plenty of time for the children to think about it visually with the help of base ten pieces.

Include the following language throughout this lesson: What is the difference between these collections (numbers)? How do these numbers compare? Which number is larger (smaller)? How much larger (smaller)?

The difference between two numbers can be represented very naturally with base ten pieces. In the words of our children, the difference consists of the pieces that have to be "added to the smaller collection (or removed from the larger one) to make both the same."

Teacher Tips

In this activity, many of our children find it helpful to use pieces of one color to show the two given numbers and pieces of another color to portray the difference between them.

34 Difference—Diagrams and Sketches

You Will Need
- Chapter 5, Addition and Subtraction
- base ten pieces for the overhead
- transparencies of Blacklines 8–13 (Base Ten Grid Paper)
- Number Card packets
  for each child
- base ten pieces
- grid paper and crayons

Your Lesson

(2–5 days) Begin with a story such as, "The children at Southside School were saving grocery receipts to earn a free computer for their class. During the first week, Mrs. Monarch's class brought in 187 receipts, while Mrs. Quire's class collected 243. How many more receipts did Mrs. Quire's children have than Mrs. Monarch's?"

Have your children form the minimal collections for 187 and 243 and discuss ways to estimate an answer to the above question.

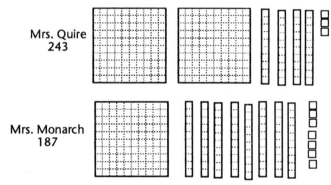

As in Insight Lesson 33, ask them to show the difference between 187 and 243 with their pieces and to then make a grid paper diagram of their work.

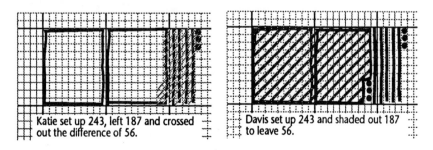

Katie set up 243, left 187 and crossed out the difference of 56.

Davis set up 243 and shaded out 187 to leave 56.

167
INSIGHT LESSONS

34 Difference—Diagrams and Sketches (continued)

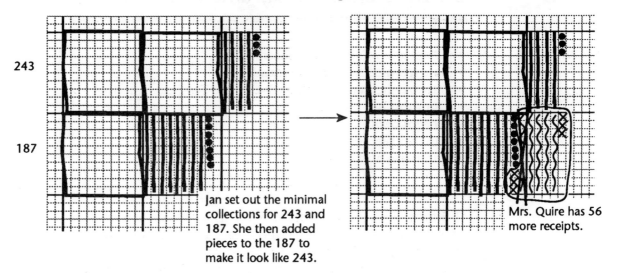

Jan set out the minimal collections for 243 and 187. She then added pieces to the 187 to make it look like 243.

Mrs. Quire has 56 more receipts.

Conduct a show-and-tell discussion of the children's thinking and diagrams.

Continue in the same manner over the next few days, changing the stories and varying the size of the numbers. When they are ready, the children will move away from the pieces and work with diagrams only. Some will even begin to use freehand sketches to show their work.

Throughout this lesson, discuss how the process of finding the difference between two numbers can be indicated in terms of both addition and subtraction. For example, here are some of the ways to indicate the difference between 243 and 187:

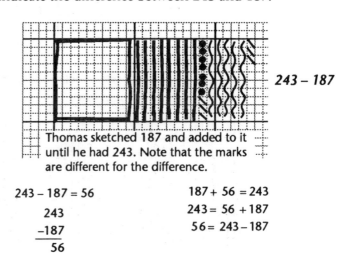

243 – 187

Thomas sketched 187 and added to it until he had 243. Note that the marks are different for the difference.

$$243 - 187 = 56 \qquad 187 + 56 = 243$$
$$\begin{array}{r}243\\-187\\\hline 56\end{array} \qquad \begin{array}{r}243 = 56 + 187\\56 = 243 - 187\end{array}$$

Encourage the children to use these operations to describe their reasoning when computing differences.

34 Difference—Diagrams and Sketches (continued)

Emphasis

When describing these stories with number operations, the symbols for addition (+) and subtraction (−) communicate the actions of combining and comparing collections of pieces, respectively.

The difference between two numbers can be thought of as the amount that is subtracted from the larger number to produce the smaller one. It is also the amount that is added to the smaller number to produce the larger one.

Practice

(Independent) Give each small group of children a packet of Number Cards. Each child can draw two numbers and diagram the difference between them. When ready, they should also express the difference in terms of number operations.

Journal Writing

Write a brief note to a friend telling them how you would find the difference between two numbers.

Homework

(Optional) Challenge your children to make diagrams or sketches of the difference in number between things like: the states east of the Mississippi and the states west of it; the estimated length of Tyrannosaurus Rex and that of Diplodocus; the number of seats in Busch Stadium and the number of seats in Riverfront stadium; the temperature on a given day in Death Valley and on that same day in Juneau, Alaska; the number of permanent teeth and primary teeth; the number of countries in South America and the number of countries in Africa; the combined ages of the boys in the class and the combined ages of the girls in the class; the number of seeds in the sunflower bloom and the number of seeds in the watermelon; the number of keys on a piano and the number of strings on a cello; the number of members of the local football team and the number of members of the local basketball team; the seating capacity of the Superdome in New Orleans and the Astrodome in Houston.

To make this even more fun, generate the list from suggestions made by the parents and children as a homework assignment.

Additional Ideas

35 Big Difference/Little Difference

You Will Need

Chapter 5, Addition and Subtraction
Version 1 and 2
- Transparencies of Blackline 95 (Double Spinner) and Blackline 96 (More or Less spinner)
- base ten grid paper and crayons or markers for each child
- base ten pieces for the overhead

Version 3
- Transparencies of Blackline 95 (Double Spinner), Blackline 97 (Triple Spinner) and Blackline 96 (More or Less spinner)
- scratch paper or calculators (if needed)

Your Lesson

(2–3 days) For additional practice with comparing numbers, have children play the following versions of Big Difference/Little Difference.

Version 1

Choose two teams and follow these directions:

Determine a beginning number by spinning the 00–90 and 0–9 spinners on Blackline 95. This number will be used by *both* teams.

Spin both spinners again to determine a second number for Team A. Spin again to determine a second number for Team B.

Each team diagrams the difference between its second number and the beginning one. These differences are compared and the More or Less spinner is used to determine the winning team.

Version 2

This is played very much like Version 1. This time, however, determine two numbers for *each* team with the spinners.

As an example: Team A spins the numbers 75 and 37 and has to find the difference between them. Then Team B spins the numbers 18 and 94 and has to find the difference between them. Team A's difference of 38 is smaller than Team B's difference of 76. Finally, the spin of the More or Less spinner determines the winner.

As you play this game, discuss the following question. Will the teams ever show the same difference? If so, what are some examples and how should the rules of the game be adjusted to handle the situation? Play and see what your children think.

35 Big Difference/Little Difference (continued)

Version 3

This is played like the other versions except that numbers are determined by spinning the 000–900, 00–90 and 0–9 spinners. With the inclusion of the 000–900 spinner on Blackline 97, some of the numbers will have 3 digits.

Teacher Tips

Find out what the children think about the question posed at the end of Version 2. It is possible for the teams to end up with the same difference. For example, the following pairs of numbers are some of the ones that have a difference of 45

    46 and 1   56 and 11   57 and 12

Your children will explore the idea of "equal differences" further in Insight Lesson 43.

These games are ones that the whole family will enjoy.

Additional Ideas

36 The Great Race

You Will Need
- Chapter 5, Addition and Subtraction
- Transparency of Blackline 95 (Double Spinner)
- a copy of Blackline 17 (The Great Race) for every two children

for each child
- base ten grid paper or scratch paper
- crayons or markers

Your Lesson

(1 day) The Great Race has the same rules as Version 1 of Big Difference/Little Difference (Insight Lesson 35), except that each team sketches its difference for each round of play on Blackline 17. Play continues until one team has sketched an accumulated total of 290–300 units with its differences. That team is then declared the winner.

Teacher Tips

Children often find it helpful to diagram strips in one place and units in the other and to use different colors for each turn.

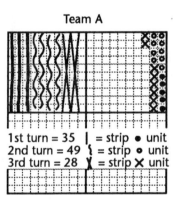
Team A
1st turn = 35 | = strip • unit
2nd turn = 49   ʆ = strip ○ unit
3rd turn = 28   ⋎ = strip ✗ unit

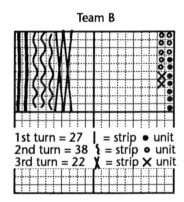
Team B
1st turn = 27 | = strip • unit
2nd turn = 38   ʆ = strip ○ unit
3rd turn = 22   ⋎ = strip ✗ unit

Additional Ideas

37 Estimating and Calculating Differences

You Will Need
- Chapter 5, Addition and Subtraction
- base ten counting pieces and grid paper as needed
- base ten pieces for the overhead
- transparencies of base ten grid paper (Blacklines 8–13)
- packets of Number Cards for each small group of children

Your Lesson

(3–5 days) In this lesson, your children will develop written records of their procedures for calculating differences.

Write the numbers 92 and 68 on the board. Divide the class into small groups and ask the groups to write a difference story that uses these two numbers, modeling it with base ten pieces or with a diagram (or sketch). Then have each group calculate the difference between 92 and 68 and write a number statement that describes their reasoning.

Groups may use materials of their choice to complete this exercise. When all are ready, have volunteers share the work of their group in a show-and-tell fashion. Allow plenty of time for this discussion — you may spend the entire day on just one problem.

Repeat the activity, having the groups begin with other pairs of numbers drawn from their Number Cards.

Throughout this lesson, encourage the children to invent (and share) their own ways of recording their thinking with number statements. They will enjoy the challenge of doing so and will come up with several methods. In all likelihood, many will discover that they do not need to show (or draw) the numbers being compared in order to complete the task. Here are examples of children's procedures that we have observed:

"In her sticker collection, Elizabeth has 92 heart stickers and 68 rainbow ones. How many more heart stickers does she have?"

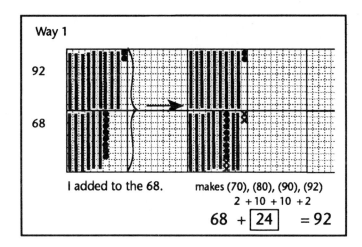

173
INSIGHT LESSONS

37 Estimating and Calculating Differences (continued)

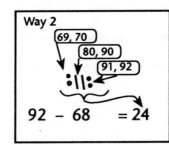
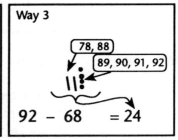

Way 4
```
  92
- 68
-----
   2  (makes 70)
  20  (makes 90)
   2  (makes 92)
-----
  24
```

Way 5
```
  92
- 68
-----
 +30  (makes 98)
 - 6  (take back 6)
-----
  24
```

Emphasis

During the show-and-tell parts of this lesson, have the children show how their number statements can be modeled with base ten pieces or pictures. This will help them clarify their thinking and, at the same time, will give you insight into what they are learning. In addition, such demonstrations are very helpful to other children. It is pleasing to see the piggybacking of ideas that takes place!

The difference between two numbers can be modeled in several ways using number pieces, drawing diagrams or sketches, or with the help of mental images. Children should have the freedom to choose from among these options, selecting the one they find most helpful in a given situation.

Encourage the children to make written records of their thinking in a manner that seems most natural to them. The "best" way is the one that is best for the child.

Practice

One way to provide practice is to use Number Cards as a source of problems. Each group can divide their packets into two piles and select a number from each pile. The group explores several ways to calculate the difference between these numbers and then moves on to another pair.

Additional Ideas

38 Differences—Teamwork

You Will Need
- Chapter 5, Addition and Subtraction
- base ten counting pieces and grid paper, as needed
- individual chalkboards, chalk and erasers
- Number Packets for each pair of children

Your Lesson

(2–3 days) In this lesson, children work in pairs to model ways for finding the difference between two numbers using base ten pieces or diagrams. They then make written descriptions of their actions. You may find it helpful to first model the activity with a volunteer from the class.

Divide the class into pairs. Each pair draws two numbers from its Number Card packet. One member of the team finds the difference between these numbers using base ten pieces or diagrams, explaining their thinking to their partner. The partner then records this method with numbers on a chalkboard.

Encourage the teams to explore a given problem in several ways before moving on to the next one. Team members trade tasks as this is done so that everyone gets experience with both parts of the activity. Rotate the Number Packets periodically.

Emphasis

For each solution, the written descriptions should reflect the way the pieces or diagrams were used and the teams should check that this has occurred.

Teacher Tips

This activity is a wonderful one for parents to do with their child!

Journal Writing

(Group) At some point, ask the group to write about one of the problems they solved. What procedures did they use? What was their thinking? How do they feel about their work?

39 Differences—Back Me Up

You Will Need
- Chapter 5, Addition and Subtraction
- base ten counting pieces and grid paper, as needed
- individual chalkboards, chalk and erasers
- Number Packets for each pair of children

Your Lesson

(2–3 days) The main activity of this lesson, Back Me Up, is a variation of that done in Insight Lesson 38. This time, however, team members may conceivably solve a problem in different ways and are challenged to explain their thinking to each other. You may wish to model it first with a volunteer from the class.

Divide the children into pairs. Each pair selects two numbers from its Number Card packet. Team members sit with their backs to one another and proceed as follows. One member of the pair (the "setter-upper") finds the difference between these numbers using base ten pieces, diagrams or sketches. The other team member (the "writer-downer") attempts to mentally picture the pieces and records a number/symbol statement to show the difference on their chalkboards.

When everyone is ready, team members compare and discuss their work. Was the correct difference found by all? Did the "setter-uppers" think about the problem in the same way as the "writer-downers"?

Continue in the same manner with children alternating tasks.

39 Differences—Back Me Up (continued)

Emphasis

"Setter-uppers" and "writer-downers" should demonstrate their thinking to each other, showing how chalkboard statements can be modeled with pieces and vice-versa. In this way, children learn to view the problems in many ways.

This activity promotes the formation of mental images for finding differences. The discussion that takes place is intended to relate these images to the use of number pieces or diagrams.

Teacher Tips

This is another activity that parents and children will enjoy trying at home. How do the adults think about these problems?

Homework

(Optional) Send each child home with several index cards and ask them to write a "difference" story on each. This will provide you with a source of problems for the next day. Distribute the cards among the teams and have them proceed to back each other up.

Additional Ideas

40 Differences—Rounding

You Will Need
- Chapter 5, Addition and Subtraction
- Number Card packets for each group of four children
- base ten number pieces and grid paper, as needed
- a copy of Blackline 16 for each child

Your Lesson

(1–2 days) To provide practice with finding differences and rounding, play Rounding Differences.

To play:

Divide the class into groups of four. Each group then separates into two teams.

In turn, have each team draw two number cards and find the difference between the numbers.

The differences are rounded to the nearest hundred and recorded in the appropriate section of the game blackline copy.

The winning team is the first to fill a column or a row, as predetermined before play.

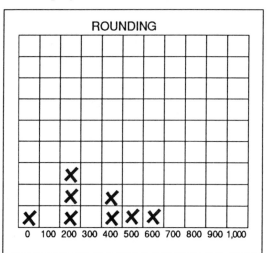

Additional Ideas

41 Calculator Differences

You Will Need
- Chapter 5, Addition and Subtraction
- base ten counting pieces and grid paper, as needed
- calculators and Number Card packets for each group of four

Your Lesson

(1–2 days) A fun way for your children to practice their estimation and calculator skills is to have them play Calculator Differences.

To play:

Within each group of four establish two teams, A and B.

The groups shuffle the cards and turn over the top card. This card is the Base Card that will be used by both teams during play.

They also decide on a margin of error for the estimates that will be made in the game. For example, one group might seek to make estimates that are within 10 of an actual answer; another might wish to come within 5.

Play then proceeds as shown in the following sample game.

1. Base Card: 204. (Estimates are to be within 10 of actual answers.)

2. Team A draws a second card: 670. Team A then estimates the difference between 670 and 204 to be 470.

3. With a calculator, Team B finds the difference between 670 and 204 to be 466.

4. Since Team A's estimate is within 10 of 466, it receives 1 point.

5. Team B now draws a second number and estimates the difference between it and 204. Team A calculates the actual difference to see if Team B receives a point.

The first team to score 10 points wins.

If time permits, additional rounds can be played, with groups having the option of changing their margin of errors. Rotate the Number Cards periodically.

Teacher Tips

How much time will be allowed for each estimate? Discuss this question in either a large or small group setting, keeping in mind the answer should be fair to all. It sometimes works well for one member of the calculating team to be a clock reader. Also, have the children decide how to handle calculating errors.

Additional Ideas

42 Differences—"Tell It All" Books

You Will Need
- Chapter 5, Addition and Subtraction
- Instructions for making "Tell It All" books (Materials Guide)

Day 1
- scrap paper

Day 2
- paper for the final copy, markers and/or crayons
- grid paper (upon request)
- a copy of Blackline 23 (Tell It All Story Book) for each child

Day 3 and 4
- chalkboard, chalk, eraser for each child
- your bound book of completed stories
- base ten grid paper or base ten counting pieces as needed (child choice)

Your Lesson

Day 1

Stories have been central to making mathematics real for your children. Now it is time for some of their stories to be "published".

Pair the children. Explain that you are beginning the process of making a "Tell-It-All" story book. Instruct each pair to write a rough draft of a difference story, to draw a visual model and then add illustrations. (See example in the Materials Guide.)

When the drafts are completed, encourage peer editing. Explain that these stories are going to be published in a book to be read by others, therefore, most language mistakes need to be corrected. Save the drafts for day 2.

Day 2

A "published" copy of the draft is created following the instructions given in the Materials Guide.

Days 3 and 4

Here it is! Today the children get to solve and enjoy stories written by their friends. As you read each story to your children, encourage reasonable estimations. After a brief discussion, ask them to model and solve the problem on their chalkboards, using a method of their choice. A related number sentence is recorded beside the model. Number statements and models are compared to those used by the authors.

When finished, place the book where it can be enjoyed by all throughout the year. Consider allowing children to "check it out" to take home for a sharing time with parents and/or interested

42 Differences—"Tell It All" Books (continued)

Teacher Tips

Illustrations help make the book more interesting and can provide visual clues to enhance the problem solving.

You or your children may think of other methods for binding the book. Try them out, too.

Publishing means sharing. When sharing, we want our work to be as correct as possible.

Additional Ideas

43 Same Differences

You Will Need
- Chapter 5, Addition and Subtraction
- base ten pieces for the overhead
- base ten pieces for each child

Your Lesson

(1–3 days) Set up the minimal collections for 11 and 15 at the overhead and ask your children to do the same at their desks. The difference between these two collections is 4. Now what happens to this difference if the following amounts are added to (or removed from) *each* collection: 1 unit; 6 units; 10 units (or a strip); 100 units (or a mat)?

The children will see that, in each case, the difference is still 4. Why is this?

Ask the children to explore further. Have them set up the minimal collections for another pair of numbers and determine the difference between the two. What happens to this difference if the same amount is added to (or removed from) *each* collection? Discuss their findings; they should discover that the difference remains unchanged.

Form the minimal collections for 42 and 28 at the overhead and ask a volunteer to show that the difference between these two numbers is 14. Ask other volunteers to show how these collections can be adjusted to show other pairs of numbers whose difference is 14. Record the numbers that are generated. See illustrations on the following page for some possibilities.

Discuss these pairs. In the opinion of the children, do any of these pairs make the difference of 14 *easy* to spot? Which pair, if any, do they prefer? For example, some may say that it is easier to picture the 14 if they use 44 and 30 instead of 42 and 28.

Repeat this discussion beginning with the minimal collections for 75 and 38. What other pairs of numbers differ by the same amount as 75 and 38? Would thinking about some of these other pairs make it easier to calculate the difference between 75 and 38?

The difference is always 4.

43 Same Differences (continued)

Discuss finding the difference between other numbers in the same way.

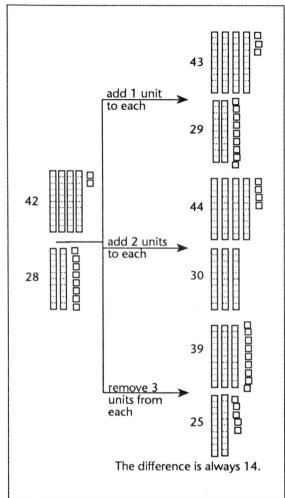

The difference is always 14.

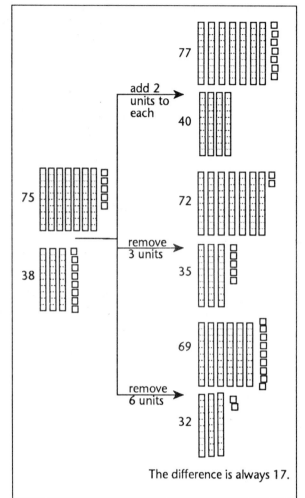

The difference is always 17.

Emphasis

The difference between two numbers remains unchanged if the same amount is added to (or subtracted from) each number. Children may sometimes use this to find the difference between two numbers.

Teacher Tips

The method for finding differences discussed in this lesson is sometimes referred to as the "equal addition" or "equal subtraction" method. It is a procedure that some children find helpful. Children differ, also, as to which pair of numbers is easiest to use.

When someone volunteers a problem for your Incredible Equations during the Calendar Extravaganza such as 54 – 36 = 18, ask if anyone can piggyback and give another problem with the same difference (perhaps 55 – 37, 64 – 46, etc.). Does anyone prefer one problem over the others? Why?

44 Difference Duel

You Will Need
- Chapter 5, Addition and Subtraction
- base ten grid paper (optional)
- Number Card packets for each pair of children

Your Lesson

(1–3 days) Have your children play Difference Duel, a challenging game that provides practice in generating pairs of numbers that have the same difference between them.

To play:

Divide the class into pairs.

Ask them to write pairs of numbers that have the same difference between them as between 42 and 27.

Team members take turns recording different pairs until one writes an incorrect pair *or* uses one that has already been used.

Which of their pairs, if any, do they think makes it especially easy to calculate the difference between 42 and 27?

When ready, the teams can repeat the game beginning with two other numbers drawn from their Number Card packets.

Journal Writing

Describe to a friend some of the ways you would use to calculate the difference between 500 and 138. Which way do you prefer?

Additional Ideas

45 Modeling Addition Stories

You Will Need
- Chapter 5, Addition and Subtraction
- base ten pieces, ten-strip boards for each child
- grid paper
- overhead base ten pieces
- Transparencies of Blacklines 8–13 (base ten grids) and Blackline 114 (ten-strip boards)

Your Lesson

PART I

(4–7 days) In this lesson, the children will model addition stories using base ten pieces, diagrams and sketches. They will also use numbers to describe their work.

Begin with a story such as, "Elayne and Bobby were collecting seashells for their grandfather. Elayne collected 53 shells and Bobby collected 44. How many shells did they have altogether?"

Conduct the following sequence of activities related to the above story. For each activity, discuss the children's work in show-and-tell fashion. Possible responses are illustrated.

Activity 1: Ask the children to set out the minimal collections for 53 and 44 (possibly on ten-strip boards). Before combining the pieces, have them estimate a reasonable answer to the story.

INSIGHT LESSONS

45 Modeling Addition Stories (continued)

Activity 2: Have the children determine the total number of shells by combining the pieces.

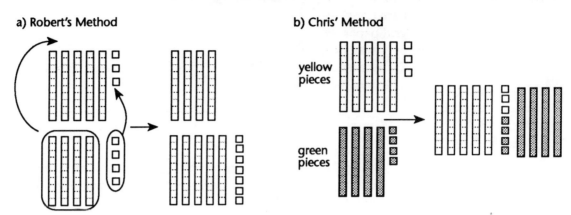

a) Robert's Method

b) Chris' Method

yellow pieces

green pieces

Note: In each case, there are 97 units in the combined collection, which represent the total number of shells.

Activity 3: Ask the children to model the story with grid paper diagrams of base ten pieces. Have them accompany each diagram with a number statement that describes their thinking.

a) Elizabeth   b) Juan   c) Seth

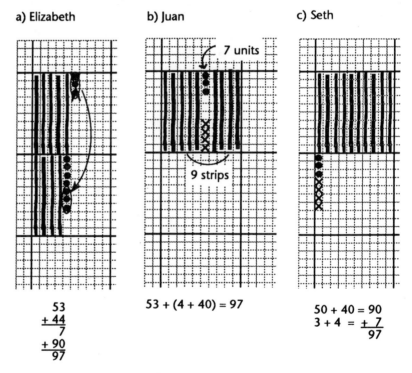

$$\begin{array}{r}53\\+44\\\hline 7\\+90\\\hline 97\end{array}$$

$53 + (4 + 40) = 97$

$$\begin{array}{r}50 + 40 = 90\\3 + 4 = +\ 7\\\hline 97\end{array}$$

Activity 4: Have the children explore the use of freehand sketches

45 Modeling Addition Stories (continued)

and numbers to model the story.

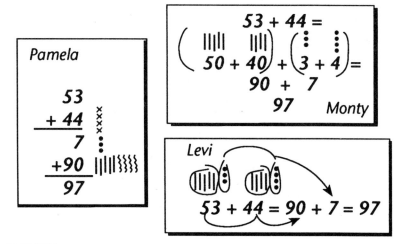

PART II

Repeat Part 1 with other addition stories. Vary the level of difficulty by including stories that use 3-digit or more addends, require regrouping, or that have unnecessary information. Note the two examples below.

During the first few days of these activities, ask the children to model each problem with pieces, diagrams and sketches. After that, encourage each child to choose the model they wish to use.

Example 1: "Mary saved 357 stamps in three years. Jose saved 226 in the same amount of time. How many stamps did they collect altogether?"

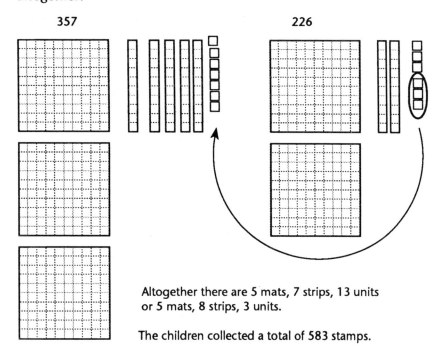

Altogether there are 5 mats, 7 strips, 13 units or 5 mats, 8 strips, 3 units.

The children collected a total of 583 stamps.

Example 2: "Ezra, Mefford and Levi walked to the fishing hole with

45 Modeling Addition Stories...(continued)

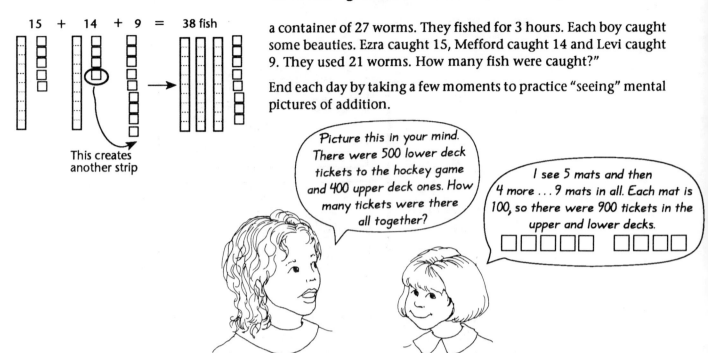

a container of 27 worms. They fished for 3 hours. Each boy caught some beauties. Ezra caught 15, Mefford caught 14 and Levi caught 9. They used 21 worms. How many fish were caught?"

End each day by taking a few moments to practice "seeing" mental pictures of addition.

Picture this in your mind. There were 500 lower deck tickets to the hockey game and 400 upper deck ones. How many tickets were there all together?

I see 5 mats and then 4 more...9 mats in all. Each mat is 100, so there were 900 tickets in the upper and lower decks.

Emphasis

Allow plenty of time for the children to show-and-tell their thinking about these problems. Don't forget to include some time for estimating.

The sum of two numbers can be modeled directly with base ten pieces, grid paper diagrams or sketches. Your children will find this to be a natural extension of their experience with base ten pieces during the place value lessons of this program.

Give the children the freedom to record procedures for addition in their own preferred style. Encourage them to describe how their model is related to the numbers they have written.

Review the use of positional notation when forming minimal collections of pieces and when reporting answers.

Teacher Tips

Addition will come naturally for your children. The base ten pieces make it possible to model both "hard" and "easy" problems in the same way, so don't be reluctant to use addends that are large or that require regrouping.

Keep the situations true to life with your storytelling. Include stories that are likely to be a part of your children's lives. Before long the children will be volunteering to tell the stories.

This is also an opportunity to integrate curriculum with stories that relate to science, social studies, etc.

45 Modeling Addition Stories (continued)

Homework

Ask each child to write some addition stories on index cards. These stories can be used as a source of problems in coming Insight Lessons. Here are samples of stories we have received:

"There were 54 scary ghosts in the haunted house. Along came 58 more. How many ghosts were haunting the house?"

"I had 66 video game tapes. How many did I have after I bought 60 more?"

"Amy has 23 coins. Chad had 39. Joe had 68. Mark had 78. How many coins in all?"

"T.J. had 1028 pieces of candy. Randy had 1100 pieces of candy. How many in all?"

"There were 87 baseball cards. I bought 62 more. How many in all?"

Additional Ideas

46 Mine, Yours, Ours—Modeling Addition

You Will Need
- Chapter 5, Addition and Subtraction
- base ten number pieces or grid paper (child choice)
- ten-strip boards (optional)

Your Lesson

(1–3 days) Divide your class into pairs and have them choose the materials they wish to use to create models of addition stories.

Tell this story: "Drew and Austin were gathering walnuts in the woods. Drew gathered 127 nuts and Austin gathered 235. How many walnuts did they gather altogether?"

Ask each team member to set up one of the addends in the story and then discuss with their partner ways to estimate and calculate the answer. Here is a description of how this might be done.

One pair, Sally and Betsy, choose to use base ten pieces to represent the two collections in the story. Sally sets up Drew's collection of walnuts (127), while Betsy shows Austin's (235).

Working together, the girls examine their pieces and estimate an answer to the question: "About how many walnuts do they have? There are 3 mats in all, so there are more than 300 walnuts. There are 5 strips and enough units to make another strip with a few left over. So we estimate that together they have between 360 and 370."

They then form the minimal collection for their combined sets of pieces to determine an exact answer to the question (see the following illustration).

46 Mine, Yours, Ours...(continued)

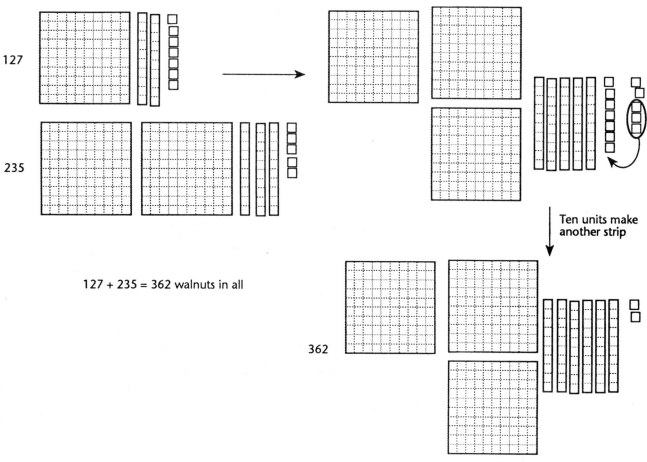

127 + 235 = 362 walnuts in all

Ten units make another strip

Conduct a show-and-tell discussion of the teams' solutions and repeat the activity with other addition stories.

Emphasis

Encourage children to work with each other when modeling stories. They will learn a lot from each other, and may find it fun, as well.

Making reasonable estimates of answers is an important component of number sense. The children will find it helpful to examine the pieces when doing this. (Does it look like there are enough units to create a strip? About how many more units are needed to make another strip? About how many strips will there be? Does it look like there are enough strips to make a mat? etc.)

Estimations are also helpful for judging the reasonableness of an answer. In today's calculator age, this is a helpful ability.

It is often helpful to think in terms of "ten-ness" when modeling addition stories (10 units make a strip; 10 strips make a mat; etc.).

Journal Writing

(Group activity) Ask the teams to describe how they modeled one of their stories and to report how they feel about their work.

47 Practicing Addition

You Will Need
- Chapter 5, Addition and Subtraction
- Transparency of Blackline 97 (Triple Spinner)
- base ten pieces or grid paper, as needed (child choice)
- Transparency of Blackline 96 (More or Less Spinner)

Your Lesson

(1–2 days) In this game, Add'em Up, the children will gain further practice with adding numbers. It takes place in a large group setting.

Divide your children into two teams and determine the team to go first.

Each team, in turn, determines two numbers to be added by spinning the spinners on Blackline 97 twice. Team members use base ten pieces or drawings to represent the addends and to find the required sum. Solutions are shared in show-and-tell fashion.

The teams determine which sum is larger. The winner of each round is determined by the spin of the More or Less spinner and is awarded a point. The round is repeated if both sums are the same.

At the end of play, the More or Less spinner is used once more to determine if the team with more (or less) points is the overall winner.

Additional Ideas

48 Addition—Phoney Numbers

You Will Need
- a copy of Blackline 18 (Phoney Numbers) for each child

Your Lesson

(1–2 days) Phoney Numbers is an activity involving diagramming and adding numbers formed from the digits of telephone numbers.

Easier Version

Each child identifies the last two digits of their home phone number. (Have a few extra numbers handy in case they are needed. We use the school number, our own number or make up a number that we pretend belongs to a favorite child performer.)

Each child chooses a partner. In turn, each partner diagrams in one section of the blackline the number represented by the last two digits of their telephone number. A Phoney Number is generated by finding the sum of these pairs of digits.

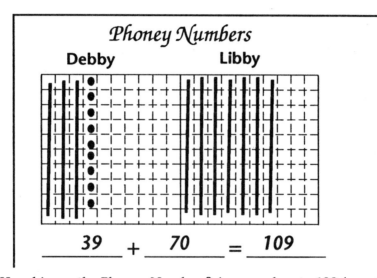

How big are the Phoney Numbers? Are any close to 100 (a mat)?

As time permits, the children rotate partners and create some more Phoney Numbers. Our children love to make a Phoney Number with each child in the class.

Harder Version

In this version, the children record the last three or four digits of their telephone numbers (as predetermined before beginning) on chalkboard or paper. They make a freehand sketch of the minimal collection for each number and proceed to add the numbers. Adjustments are made to the sketches as the minimal collection for the required sum is formed.

Now, on to a different friend to create a new Phoney Number!

48 Addition—Phoney Numbers (continued)

Teacher Tips Ask the children to suggest other ways to create Phoney Numbers. They may suggest diagramming license tag numbers, the numbers of pages in a collection of books, etc.

Journal Writing Complete this sentence: Today I enjoyed _____.

Additional Ideas

49 Jersey Day—Multiple Addends

You will need
- Chapter 5, Addition and Subtraction
- base ten number pieces or grid paper, as needed (child choice)

Your Lesson

(1 day) Plan a "Jersey Day" with your children! For the big event, ask each child to wear a sport-style jersey bearing a number. What fun your children will have adding the numbers on their jerseys! Challenge them to add 2, 3, 4 or maybe even the whole class' numbers to arrive at some "Team Numbers". What would be the Team Number for all jerseys with odd numbers; with a 3 in the 10's place; worn by children with blue eyes; with numbers written in black?

Throughout this activity, encourage the children to choose how they wish to calculate the required additions. Some may use number pieces to assist them; others may use diagrams, sketches or only numbers.

In case some children forget their jerseys, have a few extra available or else let them pick any number, write it on paper and tape it to themselves.

INSIGHT LESSONS

50 Number Card Addition

You Will Need
- Chapter 5, Addition and Subtraction
- a packet of Number Cards for every pair of children
- base ten pieces, grid paper and/or ten-strip boards, as needed (child choice)
- Blacklines 19–21 (Number Card Addition)

Your Lesson

(2–5 days) Distribute a packet of Number Cards to each pair of children. The pairs shuffle their cards and each child draws a number. The pairs work together, using the materials of their choice, to *estimate the sum* of the drawn numbers and to then *determine the exact total*. Results are recorded on Blackline 19.

The cards are discarded and the children repeat the activity with new cards.

As you walk among your children, conduct a conversation with each team, inviting them to describe their work to you.

Continue this activity for as many days as you choose—in fact, this is one you will likely revisit often.

Emphasis

Encourage the teams to explore different ways to perform the addition and to compare estimates with exact calculations. Hopefully, estimation skills are getting sharper with practice.

Teacher Tips

We usually begin with pairs of children, so just two numbers are added and then move on to larger teams. In that case, Blacklines 20 or 21 will be used.

Journal Writing

(Group activity) At some point during the lesson, ask each team to describe how they performed a particular addition. Have them include diagrams or sketches in their descriptions and share feelings about their work.

What stories can the teams tell to match some of their additions?

Assessment

The conversations that you have with the teams, together with a review of their journals, should help you gain insight into their understanding of addition concepts. You might also ask your children to make a self-assessment of their work.

Additional Ideas

51 Subtraction Take-Away Stories

You Will Need
- Chapter 5, Addition and Subtraction
- base ten number pieces, grid paper and/or ten-strip boards, as needed (child choice)
- base ten pieces
- Transparencies of Blacklines 8–13 (base ten grid) and Blackline 14 (ten-strip board)

Your Lesson

(3–7 days) Have you noticed how your children perk-up and listen when you begin to tell your math stories? Well, now it is the time for some *take-away* stories.

Begin with a story such as: "Tara and Sara were baking butter cookies to surprise their mother on her birthday. They worked and worked all afternoon and made a tremendous mess! When they were finished, there were 36 beautiful golden-brown cookies on a platter covered with plastic wrap and topped with a beautiful pink and purple polka-dotted bow.

"Since they got batter all over their clothes, they ran upstairs to change before their mother got home. While they were changing, their little brother, Leroy, ate 9 of the cookies. How many of the 36 cookies were left?"

Ask your children to model the above story by completing the following activities and, for each activity, conduct a show-and-tell discussion of the children's work. (See below for possible responses).

Have the children each choose a material for their model. They will then use base ten pieces, grid paper diagrams or freehand sketches to picture the 36 cookies that were made. Next, they will estimate the remainder by determining the reasonable answer.

Carmello's method	Melony's method
36 − 9	36 − 9 =

Carmello: Nine is very close to 10. If I remove 10 from 36, I will have 26 remaining. There were *about* 26 cookies left on the plate.

Melony: If I take away the 6 units, I will still need to take away a few more. I will get those from one of the strips. The answer should be a little less than 3 strips—or a little less than 30 units. So there were a few less than 30 cookies left on the plate.

The children will then take away 9 units to find the exact number of cookies that were left. Accompany this action with a number statement that describes it.

51 Subtraction Take-Away Stories (continued)

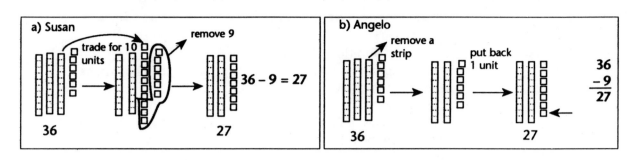

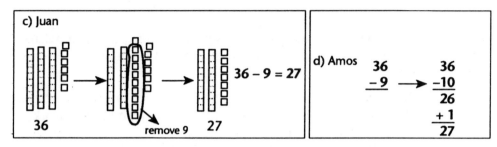

Repeat the above activity with other take-away stories. Vary the level of difficulty by including stories that require regrouping as well as some that don't, for example:

Problem 1: Opal's mother drove 247 miles in two days. On the first day, she drove 189 miles. How many miles did she travel on the second day?

Anika's method

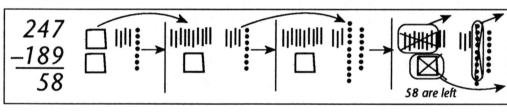

Stephan's method

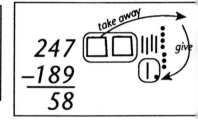

Problem 2: Mrs. Head had $345 in the bank and withdrew $123. How much money did she have left in the bank?

Peter's method

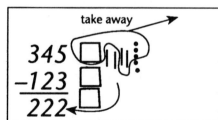

51 Subtraction Take-Away Stories...(continued)

Include stories that contain added or unnecessary information.

There were 627 children waiting in the bus room. There were also 38 children playing in the gym. After a while 125 children left the bus room on the first buses. Then 236 children left the bus room on the second buses. How many children remained in the bus room?

Dinah's method

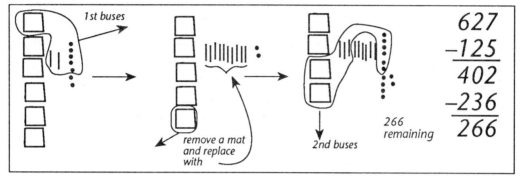

Add stories that require several steps in order to be solved.

Mr. Walls had 126 balloons in the gym. The first class popped 30 and the second class popped 25. How many balloons were left?

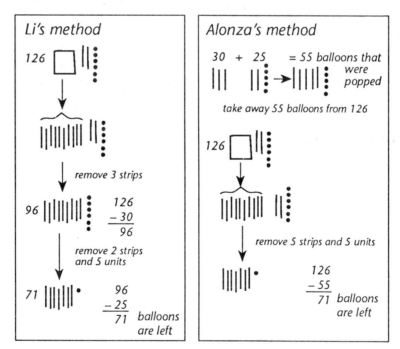

End your class each day by taking a few moments to practice "seeing" mental pictures of these problems.

51 Subtraction Take-Away Stories...(continued)

Emphasis

Modeling take-away subtraction problems with base ten pieces or diagrams is very direct. Create the minimal collection for the original amount and remove the required number of units. The collection of pieces that remains represents the answer. Sometimes pieces will need to be traded in order to complete this procedure. However, as with addition, children will find this to be a natural extension of their previous experience with counting pieces.

Give the children the freedom to record procedures for subtraction in their own style. Encourage them to describe how their use of base ten pieces corresponds to the numbers they have written.

Review the use of positional notation when forming minimal collections and when reporting answers.

When modeling take-away problems, it is sometimes necessary to work with collections that are not minimal. This happens whenever pieces need to be exchanged in order to take away the required number of units.

Teacher Tips

The base ten pieces make it possible to model both "hard" and "easy" subtraction problems in the same way, so don't be reluctant to use large numbers or stories that require regrouping. You will be pleased how children are able to picture problems that involve borrowing.

It is very important that you continue to encourage and nurture each child in their choice of ways to create a model. Success develops naturally through ownership.

Additional Ideas

52 Strategy—A Game of Take Away

You Will Need
- Chapter 5, Addition and Subtraction
- base ten pieces for each pair of children
- base ten pieces for the overhead

Your Lesson

(2–3 days) Strategy, a game played with two teams, involves removing units from a collection of base ten pieces. The following dialogue illustrates how to play this game. Here the teacher and the children are on opposing teams.

TEACHER *We are going to play Strategy. Before we begin, we must decide the number of units that can be removed in one turn.*

CHILDREN *Let's say that a team can remove no more than 9 and no less than 4 in one turn. This time, let's make the loser be the team that cannot make a move.*

> How many units can be removed?
> No more than 9   No less than 4
> **Who loses?**
> The team that can't make a move.

TEACHER *Okay, let's begin.*

The teacher places a random collection of base ten pieces on the overhead and the total number of units represented is determined.

TEACHER *How many units are represented by this collection?*

CHILDREN *10...20...34.*

In this particular game, play proceeds as shown:

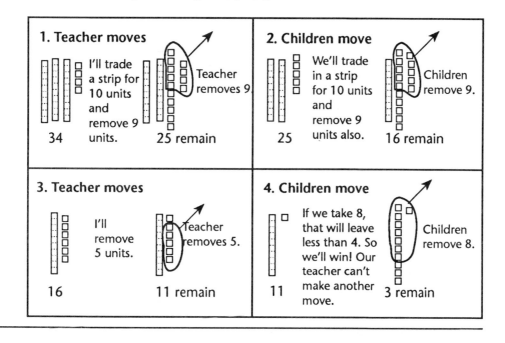

1. **Teacher moves**
   I'll trade a strip for 10 units and remove 9 units.
   34 → Teacher removes 9. 25 remain

2. **Children move**
   We'll trade in a strip for 10 units and remove 9 units also.
   25 → Children remove 9. 16 remain

3. **Teacher moves**
   I'll remove 5 units.
   16 → Teacher removes 5. 11 remain

4. **Children move**
   If we take 8, that will leave less than 4. So we'll win! Our teacher can't make another move.
   11 → Children remove 8. 3 remain

52 Strategy—A Game of Take Away (continued)

The students win because the teacher is left without a move, since, by the agreed upon rules, no less than 4 units can be removed in a turn and one loses if unable to make another move.

Teacher Tips Once your children are comfortable with the large group version of this game, turn them loose and have them play in pairs. Can anyone discover "winning" strategies to follow while playing?

Homework Encourage your children to play Strategy with their parents. Base ten pieces (Blackline 7) duplicated on construction paper can be used for this purpose.

Additional Ideas

53 Number Card Take Away

You Will Need
- Chapter 5, Addition and Subtraction
- a packet of Number Cards for every two children
- base ten pieces, grid paper and/or ten-strip boards, as needed (child choice)
- Blackline 22 (Number Card Take Away)

Your Lesson

(2–5 days) This activity is the take-away version of Number Card Addition (Insight Lesson 52).

Distribute Number Cards and Number Card Take Away sheets to each pair. The teams shuffle the cards and draw two numbers. The numbers are placed horizontally or vertically, with the larger one to the left or on top, respectively.

| 745 | 163 |

The teams work cooperatively, using materials of their choice, to set up (or picture) the larger number. They agree on an estimate of what will remain after the smaller amount is taken away from the larger, and record that estimate on Blackline 22.

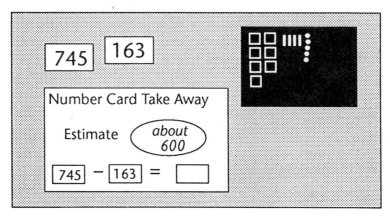

Luis: If 163 is taken away from 745, there will be about 600 left. That's because there are 7 mats and 1 will be taken away. I'll write 600 as my estimate.

Marion: I think there will be a little less than 600 because we'll have to make a trade since there aren't enough strips and units to remove 63 right away.

53 Number Card Take Away (continued)

The children proceed to calculate the exact amount that is left and record that on their blacklines, comparing it with their estimate.

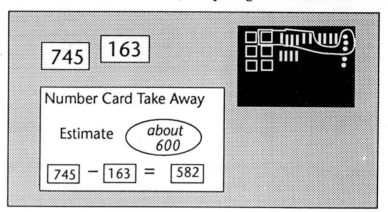

Luis: Let's trade a mat for 10 strips. Then we can remove **1 mat, 6 strips** and **3 units**.

Marion: Yes. That will leave 5 mats, 8 strips and 2 units.

Conduct conversations with the teams throughout this activity, asking them to describe their work to you. *How* did they make their estimate and *why* do they think it is reasonable? *What* method(s) did they follow to arrive at an exact answer?

As time permits, teams can draw two more cards and repeat the activity.

Continue this activity for as many days as you choose. Distribute the number card packets randomly to provide a greater variety of problems.

Emphasis

Encourage the teams to explore different ways to perform the subtractions and to compare estimates with exact calculations.

Some problems can be easily solved by pausing a moment to examine the situation before reacting.

Teacher Tips

This activity provides such great practice you will find it beneficial to reuse often.

What stories can the children tell to match the exposed number cards? Invite them to share the stories with you as you circulate and conduct conversations.

(Assessment) The conversations you have with the teams, together with a review of their journals, should help you gain insight into their understanding of this form of subtraction.

53 Number Card Take Away (continued)

Journal Writing

(Group activity) Ask the teams to select two numbers and describe how they would determine the amount left when the smaller number is taken away from the larger. Have them include diagrams or sketches in their descriptions and share feelings about their work.

What stories can the teams tell to match some of their subtractions?

Additional Ideas

54 Stories That Mix Addition and Subtraction

You Will Need
- Chapter 5, Addition and Subtraction
- overhead base ten pieces
- Transparencies of Blacklines 8–13 (base ten grid)
- have available for child choice: base ten counting pieces; ten-strip boards; base ten grid paper; pencil/paper; chalkboards, chalk and erasers

Your Lesson

(2–5 days) It is now time to begin varying the stories you tell to include both addition and subtraction (take-away and difference) problems.

For each story, ask your children to: Model the problem that is posed with number pieces or with a diagram or sketch.

Use their model to help estimate and calculate an answer to the problem.

Write a number/symbol statement that describes their thinking about the problem.

Share their solutions in a show-and-tell fashion.

Tell stories of their own. (You'll find it to be great fun—their stories are always so-o-o-o-o good!)

Emphasis

Ask the children to discuss how they decided whether to add or subtract.

There are many appropriate ways to model a problem. Your children will select the one that is most comfortable for them.

Encourage the children to show how their models correspond to the number/symbol statements they wrote.

Determining reasonable answers involves examining a situation and using common sense as well as number sense to solve the problem.

The sharing of methods challenges and tempts children to try the methods used by others. Sometimes one person's method becomes the preferred method of peers.

Journal Writing

Complete this sentence: Today I was pleased that I _____.

Additional Ideas

55 Addition and Subtraction— Teamwork

You Will Need
- Chapter 5, Addition and Subtraction
- base ten counting pieces and/or grid paper, as needed (child choice)
- individual chalk boards, chalk and erasers

Your Lesson

(2–3 days) In this lesson your children will work in teams to solve addition and subtraction problems with both models and number/symbol statements.

Divide your children into pairs. One member of each team is to *model and solve* (with pieces or pictures) the addition and subtraction story you tell. The other member uses *number statements* to describe the first person's solution. When finished, the number statement and the manipulation of pieces or pictures must reflect the same process for solving the problem.

Have the teams show-and-tell their work and then ask the children to trade tasks as you tell the next story.

Emphasis

This activity encourages the children to look at a problem from many points of view and to use numbers to describe their actions. Working with a partner and listening to others' solutions increases the chances for understanding how to model and solve problems.

Teacher Tips

Encourage your children to model the problems in their own ways, using materials of their choice. One partner may choose to use grid paper while another may simply draw free hand sketches.

Alternating jobs helps children begin to replace the counting pieces with *mental images* of the pieces—thus becoming more independent and adept with mental arithmetic.

55 Addition and Subtraction—Teamwork (continued)

Practice

(Independent) On the second or third day, list several addition or subtraction problems on the board for your children to solve in pairs, as described above. Include 3-digit numbers and problems that involve regrouping. Each team works at a comfortable pace. No two teams are compared or expected to solve the same number of problems.

This is an especially good opportunity for you to observe the children at work. Conduct conversations with each team and ask them to describe the process used to arrive at their solution. Are the children working well together? Do they seem to understand their work?

Journal Writing

(Group activity) Ask each team to write a descriptive paragraph about one of their problems. Have them comment on how they solved the problem and how their use of numbers corresponds to their model.

Assessment

Having a checklist identifying the qualities you would like to see the children exhibiting in this activity might be helpful as you converse with each team.

Additional Ideas

56 Addition and Subtraction—Tell It All Story Books

Your Lesson

Ask your children to create a Tell It All book of addition and subtraction stories. Follow the procedures described in Insight Lesson 42.

Day 1

Make rough drafts of books.

Day 2

Make final copies of books and illustrate them.

Day 3 and 4

Show and enjoy books.

Here is an example of such a story.

Additional Ideas

57 Relating Addition and Subtraction— Building Partitions

You Will Need
- Chapter 5, Addition and Subtraction
- base ten grid paper and crayons or markers

Your Lesson

(3–5 days) In buildings, partitions are used to separate a space into smaller sections. In mathematics, a collection of units can be partitioned into two smaller sets (see the first illustration below). Such partitions are helpful for demonstrating how the operations of addition and subtraction are related.

Draw a sketch of 6 strips and 2 units and then adjust it to depict the actions shown here:

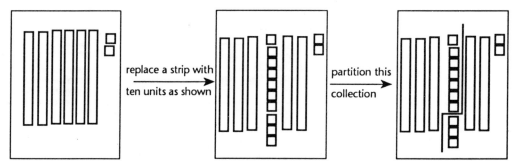

Explain that you have *partitioned* the 62 units into 2 smaller sets. Invite volunteers to write number statements that describe any addition or subtraction they see (some may want to rotate the picture as they do so).

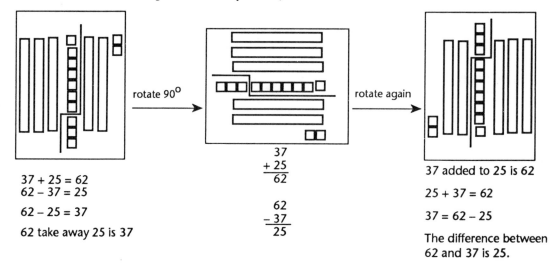

37 + 25 = 62
62 − 37 = 25
62 − 25 = 37
62 take away 25 is 37

$$\begin{array}{r}37\\+\ 25\\\hline 62\end{array}$$

$$\begin{array}{r}62\\-\ 37\\\hline 25\end{array}$$

37 added to 25 is 62

25 + 37 = 62

37 = 62 − 25

The difference between 62 and 37 is 25.

Discuss how these pictures show that addition and subtraction are related and that both operations can be seen in the same picture.

57 Relating Addition and Subtraction...(continued)

Ask children to draw a diagram of the minimal collection for 74 units on grid paper. Then, have them partition their diagram into sets of 39 and 35 units and to write addition and subtraction statements that are shown by this partition.

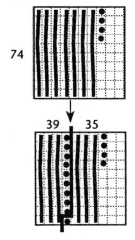

"If Mrs. Mead's class has 35 children, and Mrs. Pollett's had 39, how many children do they have together?"

39 + 35 = 74 children

"If I had 74 balloons and 35 of them flew away, how many would I have left?"

```
  74        74 − 35 = 39
− 35
  39
```

"Before any balloons flew away, there were 39 and 35 together."

Have them show-and-tell about their work, explaining what the partition shows.

Challenge them to show partitions of other sets of units, including 3-digit amounts. For example: Partition 442 units into two parts, with one part having 285 units. How many units are in the other part? What addition and subtraction statements can you write about this partition? One way of doing this:

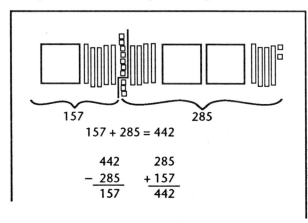

157 + 285 = 442

```
  442       285
− 285      +157
  157       442
```

57 Relating Addition and Subtraction...(continued)

Take the challenge a step further by asking your children to imagine mental pictures of partitions.

Emphasis	Partitions are a visual way of showing how addition and subtraction are related. Ask the children to describe this relationship in their own words. Encourage them to describe partitions in terms of both number and verbal statements.
Practice	(Independent) Ask children to partition a number of units any way they wish and to write related number and verbal statements that describe the partition. Some of their work can be recorded as journal entries.
Homework	Give the children a number or two to partition for homework. For example: Your total number of units is 95. Partition the 95 anywhere you choose and write a related number/symbol as well as word statements to describe what is shown by the partition.

Additional Ideas

58 Partition Challenge

You Will Need
- Chapter 5, Addition and Subtraction
- base ten grid paper and crayons, as needed

Your Lesson

(3–5 days) Divide your class into twos and have each pair follow these steps:

Select any number 100 or less to partition.

The first member draws a partition of the chosen number and records a related addition and subtraction problem about it. The other member checks this work.

Repeat the previous step by having the second member of the pair draw and describe a new partition of the same number.

Repeat until a specific partition is used twice (in any form) or until a mistake has been identified. At that time, the team begins again by choosing another number to partition.

This team is playing to partition 100.

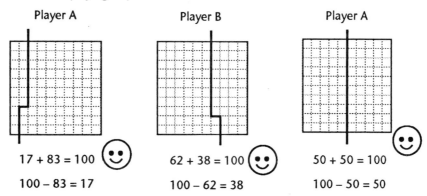

Additional Ideas

59 Ways to Partition 100

You Will Need
- Chapter 5, Addition and Subtraction
- 15 copies of Blackline 24 (Partitioning 100) for each group of four children
- markers and paper available for making books or banners

Your Lesson
(2–3 days) Divide the class into groups of four and pose the challenge of displaying all the ways to partition a collection of 100 units. Encourage the groups to make a plan for their work and to decide how they will prepare their displays. They may wish to make banners, books or posters, for example.

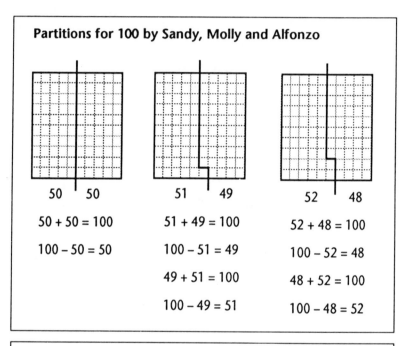

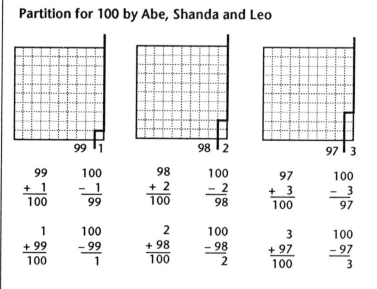

59 Ways to Partition 100 (continued)

When ready, ask the groups to describe any patterns they observe in their partitions and to comment about what they learned from the activity.

Conduct a show-and-tell discussion of the displays. Encourage children to notice the observations of others on their group's work.

Emphasis

The groups will find it helpful to organize their work. For example, they may discuss how to arrange their partitions. Should they do so in any special order? Should partitions like those shown here be counted once or twice? Etc.

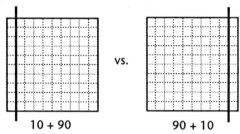

Are these partitions of 100 the same or different?

Your children may observe several patterns in their partitions. Being aware of such patterns is an important component of sound number sense.

Teacher Tips

Partitioning 100 units is an activity that your children will encounter again when they explore decimals and percents.

Additional Ideas

60 Back Me Up—Mental Pictures of Addition and Subtraction

You Will Need
- Chapter 5, Addition and Subtraction
- base ten counting pieces and grid paper, as needed
- individual chalkboards, chalk and erasers
- Number Card packets for each pair of children

Your Lesson

(3–5 days) The main activity of this lesson is structured in the same way as that done in Insight Lesson 39. Its purpose is to encourage children to explain their thinking to one another about addition or subtraction problems.

Divide the class into pairs, who sit with their backs together. Tell an addition or subtraction story. The "setter-uppers" model the story and its answer using number pieces or pictures. The "writer-downers" attempt to mentally picture the problem and use numbers and symbols to compute an answer. When all team members are ready, they compare and discuss their work.

"Jennifer's Mom brought 40 cookies to a party and her friends ate 28 of them. How many cookies did she have left to take home?"

Continue in the same way with other stories, having the team members trade tasks each time.

Once it appears that teams are ready to move at their own pace, ask the children to use Number Cards as a source for problems. For each problem, a team draws two cards and decides if it wishes to add or subtract them (or even do both).

Encourage your children to describe their work as you circulate among them.

60 Back Me Up—Mental Pictures...(continued)

Emphasis

"Setter-uppers" and "writer-downers" should demonstrate their thinking to each other, showing how chalkboard statements can be modeled with pieces and vice-versa. In this way, children learn to view the problems in many ways.

This activity *promotes* the formation of mental images for performing addition and subtraction. The discussion that occurs is intended to relate these images to the use of number pieces or diagrams.

Assessment

Develop a checklist to use as you observe and talk with your children during this activity.

Homework

(Optional) Send each child home with several index cards and ask them to write an addition or subtraction story on each. This will provide you with a source of problems for the next day. Distribute the cards among the teams and have them proceed to "back" each other up.

Additional Ideas

61 Take A Long Hard Look—Mental Arithmetic and Calculating Options

You Will Need
- Chapter 5, Addition and Subtraction
- base ten counting pieces for the overhead
- base ten counting pieces for each child
- Number Card packets for every two children
- base ten grid paper and calculators, as needed

Your Lesson

(2–4 days) Write the problem 128 + 99 on the overhead. Have the children form the minimal collection for each number with their pieces and discuss ways to perform the required addition. In particular, challenge them to use mental arithmetic to compute an answer. Here are some possible responses:

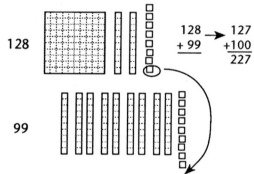

"I imagined making a mat by moving a unit from 128 to 99. That makes 127 + 100 or 227."

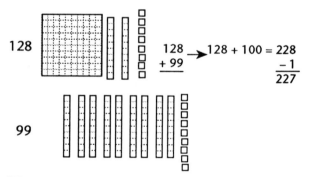

"The 99 is almost a mat, so I added 128 and 100 and then took away 1."

61 Take A Long Hard Look...(continued)

Repeat this activity for the following sums: 78 + 44; 253 + 75

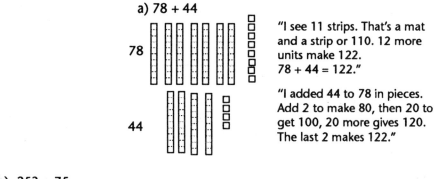

a) 78 + 44

78

44

"I see 11 strips. That's a mat and a strip or 110. 12 more units make 122.
78 + 44 = 122."

"I added 44 to 78 in pieces. Add 2 to make 80, then 20 to get 100, 20 more gives 120. The last 2 makes 122."

b) 253 + 75

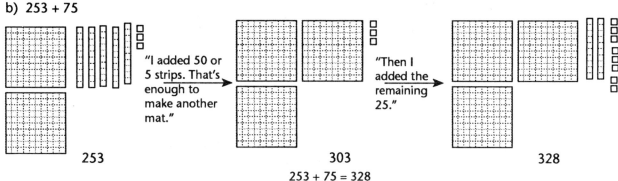

"I added 50 or 5 strips. That's enough to make another mat."

"Then I added the remaining 25."

253   303   328

253 + 75 = 328

Arrange your children in pairs. Have each team draw two Number Cards and then *take a long hard look* at the numbers.

How would they add the numbers?

Could they adjust the numbers in any way so as to make their task easier? If so, can they perform the addition mentally, perhaps with the help of a sketch? If not, do they feel a calculator is needed?

Have them write a sentence or two that describes their thinking about the problem. When finished, they can then select another set of numbers to add.

Carry on conversations with the teams. For what pairs of numbers were they able to make a "mental image" adjustment? Were there any pairs for which they weren't able to make such an adjustment? Were there any additions for which it seemed most appropriate to use a calculator? Why?

Repeat the above activities, only this time exploring subtraction problems. You will be surprised at the way some children adjust numbers to make subtraction easier.

61 Take A Long Hard Look...(continued)

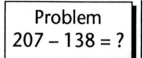

Problem
207 − 138 = ?

Strategy 1

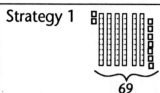

"I imagined adding 2 to 138 to make 140. Then 60 more makes 200. The last 7 gives 207. 207 − 138 = 69."

Strategy 2

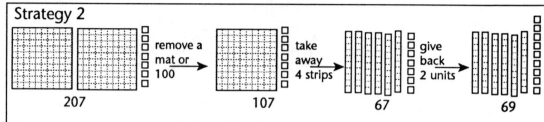

207 → remove a mat or 100 → 107 → take away 4 strips → 67 → give back 2 units → 69

"I thought of 207 and imagined taking away a mat—that left 107. I needed to take away 38 more, so I removed 4 strips (40) and gave back 2 units."

207 − 138 = 207 − 100 − 40 + 2 = 69

Emphasis

A major purpose of this lesson is to encourage children to select the calculating option(s) that seems to be appropriate for a computation. Encourage them to look for opportunities to use mental arithmetic, but recognize there are computations for which a calculator is a reasonable choice. Children will vary in the option they pick, depending on how they think about the numbers.

The process an individual uses when performing a mental calculation is very personal. Children enjoy seeing how others do the problems and learn much from the discussion.

Various adjustments may be suggested by the children. As long as the adjustment works for the user, it is acceptable. For example, when subtracting 80 from 264, a child may suggest removing 100, resulting in 164; and then giving back 20 making a total of 184.

Teacher Tips

Rotate the Number Card packets throughout this lesson. Some children will notice adjustments that will knock your socks off! Others will not see those so obvious to you. Accept and enjoy these differences in your children.

Journal Writing

How do you feel when doing addition and subtraction "in your head"? What is hard for you? What is easy?

Additional Ideas

62 Estimation, Rounding and Calculators

You Will Need
- Chapter 5, Addition and Subtraction
- Blackline 16 (Rounding) for each child
- Number Card packet and calculator for every two children
- base ten number pieces and grid paper, as needed

Your Lesson

(2–3 days) During this lesson, your children will strengthen their estimation, rounding, graphing and calculator skills by playing the game of Rounding Graphs.

To play the game:

Divide the class into pairs and distribute the materials listed above. The children within each pair oppose one another during play. Prior to beginning, opponents decide who plays first and whether they must complete a row or a column of Blackline 16 to win.

Each pair of opponents shuffles its packet of Number Cards and splits it into two piles, face down. The top card from each pile is turned over.

Player A decides whether to add the numbers on the cards or to subtract the smaller number from the larger. Player A then estimates this answer and rounds the estimate to the nearest hundred. The rounded value is reported to Player B.

Player B uses the calculator to perform the same calculation. If Player A's estimate was accurate, it is recorded in the appropriate column of A's record sheet. If the guess does not round off as predicted, the turn is forfeited.

Opponents then shift roles and the process is repeated with two new numbers. Play continues until one player has completely filled a column or row (as predetermined) on their blackline.

62 Estimation, Rounding and Calculators (continued)

Teacher Tips This game is a favorite—to be revisited often—for fun and practice!

Some children may wish to use base ten pieces, grid paper or free-hand sketches to help them with their estimates.

You may wish to teach this game to your class by first playing teacher vs. class.

Additional Ideas

63 Number Card Tic-Tac-Toe—Mental Arithmetic and Problem Solving

You Will Need
- Chapter 5, Addition and Subtraction
- a packet of Number Cards and a calculator for every two children
- base ten counting pieces and grid paper, as needed

Your Lesson

(2–4 days) Ask the children to choose a partner. Distribute the materials to each pair of children. In turn, the children prepare a tic-tac-toe grid for play by writing in any multiple of 100 (from 100 to 2,000). Multiples can be used more that once if so desired.

600	800	1500
1400	100	200
700	500	300

Players decide who will go first and then follow these steps to play:

Ten Number Cards are randomly selected and placed face up on the playing surface.

Player A inspects the playing grid, chooses a block and picks two cards to add *or* subtract in order to claim the desired space.

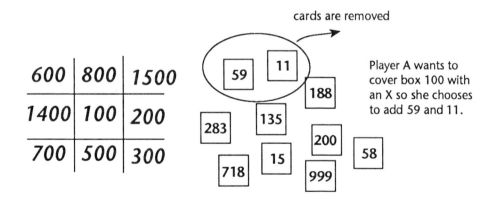

Player A wants to cover box 100 with an X so she chooses to add 59 and 11.

The problem is solved using mental arithmetic strategies. The solution is rounded off to the nearest hundred. The problem is checked on the calculator by Player B and the move is approved or disapproved. If the move is approved, Player A records in the block both the problem and the X. If the problem is disapproved, play goes to Player B.

63 Number Card Tic-Tac-Toe...(continued)

600	800	1500
1400	~~100~~ 59 + 11	200
700	500	300

The two Number Cards are replaced by new ones and the process is repeated with the players trading roles.

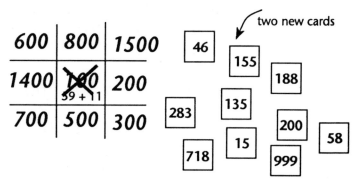

Play continues until a player is able to claim three blocks in a row (horizontally, vertically or diagonally) or until a draw is reached.

As time permits, additional rounds are played using other grids.

Children may use number pieces or grid paper to help with their mental calculations.

Additional Ideas

64 How'd Ja Do It?

You Will Need
- Chapter 5, Addition and Subtraction
- chalkboards, chalk and erasers for every two children

Your Lesson

(2 days) Divide the class into teams of two. Tell an addition or subtraction (difference and take-away) story and challenge the teams to brainstorm ways to model the story with numbers, symbols and sketches.

Circulate among the teams as they do this and conduct conversations with them about the methods they used to solve a given problem.

When they are ready, have each team member describe one of the methods in their Math Journals. Encourage them to communicate their thoughts clearly using both words and pictures.

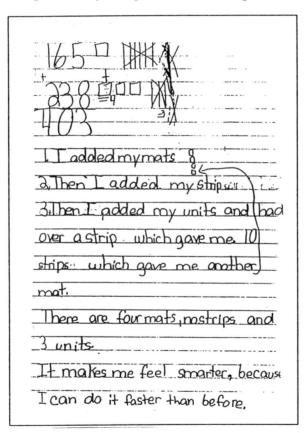

Emphasis

The main purpose of this lesson is to encourage children to solve the same problem with a variety of methods. As a result, both oral and written mathematical communication abound.

The stories can be modeled in several ways. Ask the children to describe their feelings about the method on which they previously wrote. Did they select it for any special reason?

64 How'd Ja Do It? (continued)

Teacher Tips This activity can be revisited often. For example, you may wish to have the children model a story and describe their work as you complete your morning clerical responsibilities.

Homework (Optional) Assign a problem and challenge your children to solve it and write a descriptive paragraph describing the process used. When the homework is returned, discuss the variety of methods used. Encourage the use of pictures to prove a point. Post the problem and children's paragraphs on the bulletin board. Invite the children to take some time to read the descriptions of the methods used by others.

Additional Ideas

65 Do It All

You Will Need
- Chapter 5, Addition and Subtraction
- a packet of Number Cards for every two children
- base ten counting pieces and grid paper, as needed (child choice)

Your Lesson

(1–3 days) Instruct the children to shuffle the number cards and place them face down in two equal stacks. Each child chooses one card from the top of each stack to serve as the source of the problem to be solved *independently* in this activity.

Challenge each child to use the two numbers to do the following:

Write an addition story problem, show an addition model and a related number statement.

Write a difference story problem, show a difference model and a related number statement.

Write a take-away story problem, show a take-away model and a related number statement.

Circulate among your children, observing their work and conducting individual conversations about it.

Allow time for the children to show-and-tell their work.

Emphasis

Ask the children to discuss how their problems indicate each operation.

If the difference and take-away stories involve the same two numbers, then their numerical answers will be the same. While this is true, it is helpful to notice that the way of modeling them is different. The difference model compares two sets of units; the take-away begins with a collection of units and removes some from it.

The context of the story must be clearly stated for an appropriate method to be applied.

Teacher Tips

Encourage your children to make displays of their work (some may even want to create more than one). Also, this activity seems to be a favorite among the children and you may wish to revisit it often.

Enjoy the differences displayed by your children's work. Such differences make the world a much more interesting place.

In the latter stages of this lesson, ask the children to write addition and subtraction problems that involve three or more numbers or more than one step.

65 Do It All (continued)

Journal Writing   Describe how you felt about the stories you created during this lesson.

Assessment   Would a checklist help here? You might want to create one that indicates things you'd like to know about each child's work.

Homework   (Optional) Give your children several nights to complete an activity similar to this at home.

Additional Ideas

66 Addition and Subtraction in the Mind's Eye

You Will Need
- Chapter 5, Addition and Subtraction
- paper and pencil for each child

Your Lesson

(1–2 days) Practice mental arithmetic exercises with the children. Write the following problem on the board and ask them to solve it:

What is 322 + 538?

Can they visualize this in their "mind's eye"? Do the numbers trigger mental images of pieces or sketches that help formulate an answer? Conduct a show-and-tell discussion of the children's thinking.

Do the same for another addition or subtraction that might be performed mentally. This time ask children to write a paragraph that describes their method of performing the calculation.

Ask them to include an illustration of their work and comment about such things as: What was their "mental picture"? How did this picture help solve the problem?

How did they feel about their solution (e.g., proud, sad, excited, powerful, frightened, etc.)?

Emphasis

Each of us begin to use mental pictures in our own way. Few of us ever see exactly the same image. However, most of us arrive at the same answer.

Mental pictures can be described with words—empowering the owner to reach higher levels of problem solving.

Teacher Tips

Moving away from manipulatives to "mental pictures" is a very personal thing. Encourage any child having difficulty to continue to use sketches or number pieces. We are confident, however, that all children will strengthen their mental arithmetic capabilities through the activities of this lesson. Everyone usually picks up on the ideas of others and contributes special ways of looking at problems—insuring success for every child every day.

Additional Ideas

67 Toy Store Spree

You Will Need
- several toy catalogues
- problem-solving materials available for child choice: calculators, paper, pencils, grid paper, art supplies

Your Lesson

(2–3 days) Pose this problem to your children:

"You have just won $1,000 in a cereal contest that is to be spent on toys! The money will be forfeited, however, if you make a list of toys that cost more than $1,000. What toys will you buy?"

Challenge each team of four to create a list of toys to buy, keeping a running total of how much money has been spent and how much is left. Ask the teams to prepare displays of their work, using available materials. Some may use a calculator; some may type their lists on a computer; others may paste pictures of toys on poster board. Encourage them to make a plan of their procedures and to be mindful of the limit on spending.

When finished, have each group report about its project to the class.

Teacher Tips

You may wish to change the focus of this project. Some other possibilities include: purchasing equipment for the gym; calculating the cost of a trip to the museum for the class or for the school; estimating the total number of miles driven by all the vehicles in the parking lot.

As you begin working with multiplication and division lessons, take a day or two occasionally to revisit addition and subtraction. You may wish to play some of the old favorite games or assign a project such as the one in this lesson.

Additional Ideas

Index

A

Acute Angle: *See Angle*

Addition
TRM: pp. 34-46
Vol 1: IL 12, 45-50, 54-67
Vol 3: Les B, C, 17, 47
See also: Calculating Options, Subtraction

Angle
TRM: pp. 89-93
Vol 1: CL 51
Vol 3: Les 16, 37, 40, 41, 42, 44, 45, 49

Area
TRM: pp. 62-64, 66
Vol 1: CL 67, 68, 70
Vol 2: CL 92, 102, 129, 130-134
Vol 3: Les 3, 4, 5, 9, 11, 13, 14, 15, 16, 22, 23, 30, 31, 32, 40, 41, 42, 43, 45, 49, 50
See also: Area Models for Multiplication and Division

Area Models for Multiplication and Division
TRM: pp. 49-50, 120-124
Vol 2: IL 78-100
Vol 3: Les 14, 15, Appendix B
See also: Arrays, Multiplication, Division, Problem Solving

Arrays (Rectangular)
TRM: pp. 2-5, 47-61, 115-117, 120-124
Vol 2: IL 78-100
Vol 3: Les 14, 15, Appendix B
See also: Extended Counting Patterns, Distributive Law

Assessment
Introduction, p. 8
TRM: Chapter 12
Assessment strategies are suggested throughout lessons

Average
TRM: pp. 83-85
Vol 2: IL 101-102
See also: Probability, Data Analysis, Statistics, Babylonian Numeration System
Vol 3: Les 46

B

Bases (Non-Decimal)
TRM: pp. 18-26
Vol 1: IL 5-15
Vol 3: Les C, 3, 4, 24, 31, 32, 46
See also: Place Value

Calculating Options (Mental Arithmetic, Estimation, Models, Sketches, Calculators, Paper-and-Pencil Procedures)
TRM: pp. 45, 46, 59-61, 104, 115-124
Vol 1: IL 5-67; CL 5-9, 19-24, 40-42, 44-48, 83-88
Vol 2: IL 68-102; CL 108-113, 124-128, 143-147, 174-179
Vol 3: Introduction (p. 7), Les B, C, 2, 3, 4, 5, 6, 10, 11, 13, 14, 15, 17, 22, 23, 24, 26, 27, 28, 29, 43, 46, 47, 50, 51
See also: Addition, Division, Extended Counting Patterns, Multiplication, Subtraction, Problem Solving

C

Calculators: *See Calculating Options*

Calendar Extravaganza
TRM: pp. ix, 101-128
Vol 2: CL 185

Capacity: *See Volume*

Changing the Unit
Vol 3: Les C, 30, 31, 32, 50
See also: Measurement, Unit

Circle
TRM: p. 93
Vol 1: CL 28
Vol 2: CL 137

Circumference
TRM: p. 93
Vol 1: CL 28
Vol 2: CL 137

Comparison (Difference) Model for Subtraction
TRM: pp. 42-43
Vol 1: IL 33-44
See also: Subtraction

Composite Number
TRM: pp. 59, 121, 124
Vol 2: IL 91
Vol 3: Les 22, 23

Computers:
Vol 3: Les 6, 33, 34

Cone
TRM: p. 93
See also: Geometry

Congruence
TRM: pp. 90, 93
Vol 1: CL 61, 62, 67, 68, 70
Vol 3: Les 7, 9, 11, 16, 20, 21, 37, 49
See also: Geometry

Contact Lessons (description)
TRM: pp. ix-x

Cube
TRM: p. 93
See also: Geometry, Measurement

Cylinder
TRM: p. 93
Vol 1: CL 25
See also: Geometry

Index Key: TRM – Teaching Reference Manual; CL – Contact Lesson(s); IL – Insight Lesson(s); Les – Lesson(s)

Index (continued)

D

Data Analysis, Graphing
TRM: pp. 79-85, 106-110
Vol 1: CL 2, 3, 14, 39, 56, 59, 60, 82
Vol 2: 106, 117, 135, 137, 140-142, 160-161, 180-183
Vol 3: Les 6, 18, 19, 33, 34, 44, 45, 51

Decagon
TRM: p. 93

Decimals
TRM: pp. 31-33, 118-120
Vol 2: IL 68-75
Vol 3: Les B, C, 2, 17, 24, 32, 47
See also: Fractions, Place Value, Problem Solving

Diameter
TRM: p. 93

Difference (Comparison) Model for Subtraction
TRM: pp. 42-43
Vol 1: IL 33-44
See also: Subtraction

Dimension (relating 2- and 3-dimensions)
TRM: p. 92
Vol 1: CL 70, 71, 91
Vol 2: CL 119, 121
Vol 3: Les 12, 20, 21 39

Dimensions (of a rectangle)
TRM: pp. 2-5, 47-61, 115-117, 120-124
Vol 2: IL 78-100
Vol 3: Les 7, 12, 13, 14, 15, 22, 32, 43, Appendix B
See also: Extended Counting Patterns, Division, Multiplication

Distributive Property of Multiplication
TRM: pp. 54
Vol 2: IL 94-96
Vol 3: Les 13, 14, 15, 22, Appendix B

Division
TRM: pp. 47-61
Vol 2: IL 76-100
Vol 3: Appendix B, Les B, C, 14, 15, 17, 22, 23, 24, 27, 28, 29, 50
See also: Area Models for Multiplication and Division, Arrays, Calculating Options, Extended Counting Patterns

Divisor: See Factor

Dodecagon
TRM: p. 93

E

Egyptian Numeration System
Vol 3: Les 24, Appendix C

Equilateral Triangle
TRM: p. 93
Vol 3: Les 16
See also: Fractions, Geometry

Estimation
TRM: pp. 45, 55, 59-61, 66-67, 76-77
Vol 1: IL 2, 37, 45, 46, 51, 61-63, 66; CL 11, 13, 15, 28-30, 37, 55-57, 69, 78, 89, 90
Vol 2: IL 83; CL 92-94, 97, 98, 102, 138, 156, 166, 171
Vol 3: Les 8, 11, 25, 31, 36, 48-50
See also: Calculating Options, Measurement, Probability

Even Number
TRM: pp. 59, 121
Vol 3: Les 6, 27, 51

Extended Counting Patterns
TRM: pp. 52-53, 115-117
Vol 1: CL 5-9, 19-24, 44-48, 83-88
Vol 2: IL 85; CL 108-113, 124-128, 143-147, 174-179
Vol 3: Les 27

F

Factor
TRM: pp. 49, 59, 121
Vol 1: CL 5-9, 19-24, 44-49, 83-88
Vol 2: IL 78-100; CL 108-113, 124-128, 143-147, 174-179
Vol 3: Les 14, 15, 17, 22, 23, Appendix B
See also: Dimension (of a rectangle), Area Models for Multiplication and Division

Fraction Bars
TRM: pp. 71-75, 125, 126, 129, 133
See also: Fractions

Fractions
TRM: pp. 69-75, 125-126
Vol 1: CL 31-35, 43, 58, 73-75
Vol 2: IL 68-75; CL 101, 114-116, 120, 148-153 155, 172
Vol 3: Les B, C, 3, 4, 5, 11, 17, 31, 33, 34, 38, 42, 43, 47, 48
See also: Decimals, Place Value

G

Geometry
TRM: pp. 86-94
Vol 1: CL 4, 17, 25-27, 36, 50-54, 61-63, 66-68, 70, 91
Vol 2: CL 103, 104, 118, 119, 121, 122, 154, 158, 162-165
Vol 3: Les A, 5, 7, 8, 9, 11, 12, 16, 20, 21, 30, 31, 37, 39-43, 45, 49

Graphing: See Data Analysis

H

Hexagon
TRM: p. 93
See also: Geometry

Index (continued)

I

Insight Lessons (description)
TRM: p. ix

Isosceles Triangle
TRM: p. 93
See also: Geometry

J, K

L

Length
TRM: pp. 62-66
Vol 1: CL 13, 37, 67, 68, 79
Vol 2: CL 93, 94, 133, 134, 137, 166. 167-169
Vol 3: Les 7, 8, 13, 14, 15, 22, 32, 40-43

M

Mean
TRM: p. 83
See also: Average

Measurement
TRM: pp. 62-68
Vol 1: CL 11, 28-30, 37, 43, 54-56, 69-72, 78, 79, 89, 90
Vol 2: CL 92-94, 97-99, 102, 129-134, 136, 138, 166-169, 171
Vol 3: Les 8, 11, 12, 13-16, 20, 21, 25, 30-32, 36, 39-45, 48-51
See also: Area, Estimation, Length, Perimeter, Volume, Weight

Measurement Sense
TRM: p. 62
See also: Measurement

Median
TRM: p. 83
See also: Average

Mental Arithmetic: *See Calculating Options, Estimation*

Minimal Collection
TRM: pp. 20-22, 24-27, 30-33
Vol 1: IL 9-11, 18-31
Vol 2: IL 69-73
Vol 3: Les C, 3, 4, 31, 32
See also: Place Value, Positional Notation

Mode
TRM: p. 83
See also: Average

Multiple
TRM: pp. 47-48
Vol 2: IL 76-77
See also: Extended Counting Patterns, Factor

Multiplication
TRM: pp. 47-61
Vol 2: IL 76-100
Vol 3: Appendix B, Les B, C, 14, 15, 17, 22, 23, 24, 27, 28, 29, 50
See also: Arrays, Area Model for Multiplication and Division, Calculating Options, Extended Counting Patterns, Problem Solving

N

Number Operations: *See Addition, Calculating Options, Division, Multiplication, Subtraction*

Number Sense
TRM: pp. 34-38
See also: Calculating Options

O

Obtuse Angle
TRM: p. 93
See also: Geometry

Octagon
TRM: p. 93

Odd Number
TRM: pp. 59, 121
Vol 3: Les 6, 27, 51

Operation Sense:
TRM: p. 47
See also: Calculating Options

P

Parallel Lines
TRM: pp. 89, 92
Vol 3: Les 40, 41
See also: Geometry

Parallelogram
TRM: p. 93

Pattern
TRM: pp. 11-17, 104-105, 115-117
Vol 1: IL 5, 13-16, 59, CL 5-9, 16, 19-24, 44-48, 83-88
Vol 2: IL 69, 85, 95, 96, 108-113, 124-128, 143-147, 174-179
Vol 3: Les 1, 3, 4, 13-15, 17, 26, 27, 31, 32, 47
See also: Extended Counting Patterns, Place Value

Pattern Blocks
TRM: pp. 73-74
See also: Fractions, Symmetry, Tessellations

Pentagon
TRM: p. 93

Perimeter
Vol 1: CL 79
Vol 2: CL 94, 133, 134, 166
Vol 3: Les 7, 13, 32, 40-43, 49, 51
See also: Length

Index (continued)

Perpendicular Lines
TRM: p. 93

Place Value
TRM: pp. 18-33, 105-106
Vol 1: IL 5-32; CL 15, 57
Vol 2: IL 68-70, 72; CL 102, 156
Vol 3: Les B, C, 3, 4, 13, 24, 31, 32, 50

Polygon
TRM: p. 93
See also: Geometry

Positional Notation
TRM: pp. 18-20, 27-28, 31, 105-106, 118-120
Vol 1: IL 18-21, 27, 32
Vol 2: IL 95
Vol 3: Les B, C, 4, 24, 31, 32, 46, 50
See also: Minimal Collection, Place Value

Prime Number
TRM: pp. 59, 121, 124
Vol 2: IL 91
Vol 3: Les 22, 23

Prism (rectangular)
TRM: p. 93

Probability
TRM: pp. 76-78
Vol 1: CL 14, 39, 82
Vol 2: CL 106, 140-141, 160-161 (Crossing the Mississippi)
Vol 3: Les 6, 33, 34, 44, 45, 51
See also: Data Analysis, Graphing

Problem Solving: *Opening Eyes to Mathematics* provides ongoing opportunities for children to explore mathematics and develop their problem solving (and problem posing) abilities. In most lessons, mathematical investigations are used as a context for introducing concepts and strengthening skills. Other lessons, such as those listed below, focus principally on analyzing problems related to school, home life, or game situations. *See also:* TRM: Chapter 1, Vol 3: Introduction.
Vol 1: CL 40-42, 44, 64, 65, 80, 81
Vol 2: CL 100, 105, 107, 123, 137, 139, 159, 162, 163, 173, 184
Vol 3: Les 2, 6, 10, 12, 19, 28, 29, 35

Q

Quadrilateral
TRM: p. 93
See also: Geometry

R

Radius
TRM: p. 93

Rectangle
TRM: pp. 93, 120
See also: Arrays

Reflective Symmetry
TRM: pp. 86-88

Vol 1: CL 17, 26, 61, 62, 66
Vol 3: Les 9, 37, 40, 42, 49, Appendix D
See also: Symmetry

Rhombus
TRM: p. 93
See also: Fractions, Geometry

Right Angle
TRM: p. 93
See also: Geometry

Right Triangle
TRM: p. 93
See also: Geometry

Rotational Symmetry
Vol 3: Les 37, 40, 42, 49, Appendix D
See also: Symmetry

S

Scalene Triangle
TRM: p. 93
See also: Geometry

Segment
TRM: p. 92

Similar Figures
TRM: p. 93
Vol 3: Les 40-43

Sorting
TRM: pp. 7-10
Vol 1: IL 1; CL 1, 2, 3, 12, 14, 58, 66, 76
Vol 2: CL 101, 109, 148, 155, 162, 170

Spatial Visualization: *See Geometry*

Sphere
TRM: p. 93

Square
TRM: p. 93
See also: Arrays, Fractions, Geometry, Measurement

Statistics: *See Data Analysis, Graphing*

Story Problems: *See Problem Solving*

Subtraction
TRM: pp. 34-46
Vol 1: IL 33-44, 51-67
Vol 3: Les B, C, 17, 47
See also: Addition, Calculating Options, Story Problems

Surface Area
Vol 1: CL 71, 72
Vol 3: Les 12, 20, 39

Symmetry
TRM: pp. 86-88, 90
Vol 1: CL 17, 26, 61, 62, 66
Vol 3: Les 9, 37, 40, 42, 49, Appendix D
See also: Geometry, Reflective Symmetry, Rotational Symmetry

Index (continued)

T

Tessellations
TRM: pp. 91-92
Vol 3: Les 42

Trapezoid
TRM: p. 93
See also: Fractions, Geometry

U

Unit
TRM: pp. 19-33, 62-66, 71
See also: Arrays, Addition, Division, Fractions, Measurement, Multiplication, Place Value, Subtraction

V

Venn Diagram: *See Graphing, Sorting*

Volume
TRM: pp. 62-64, 67
Vol 1: CL 29, 30, 69, 90
Vol 2: CL 97-99
Vol 3: Les 12, 20, 21, 36, 39

W

Weight
TRM: pp. 62, 63, 67
Vol 1: CL 11, 55, 56, 89
Vol 3: Les 25, 36

Y

Z

Zero
TRM: pp. 19, 20, 27, 32, 59
Vol 3: Les 24, 46, Appendix B
See also: Calculating Options, Number Operations, Place Value